Electrical Wiring & Lighting
for Home & Office

No. 671
$7.95

Electrical Wiring & Lighting for Home & Office

By Edward L. Safford

TAB BOOKS
Blue Ridge Summit, Pa. 17214

FIRST EDITION

FIRST PRINTING— DECEMBER 1973

Copyright ©1973 by TAB BOOKS

Printed in the United States
of America

Hardbound Edition: International Standard Book No. 0-8306-3671-4

Paperbound Edition: International Standard Book No. 0-8306-2671-9

Library of Congress Card Number: 73-86765

PREFACE

The magic genie of our world is electricity. It makes our modern life possible (and bearable) because of its diverse uses. Think for a moment of the thousands of ways electricity serves you, day and night. Imagine what the result would be if, suddenly, there were no electricity in the home, the car, the streets or cities—anywhere! Reminds me of a movie where a visitor from another galaxy caused all electricity in the world to become ineffective for 30 minutes. Automobiles stopped on the streets, homes became dark, radio and television vanished, factories stopped, food began to spoil, medicines lost their potency for lack of refrigeration—to see what **could** happen was quite an experience! It was reassuring to know we do have as sure a supply of this vital, life-giving phenomenon as we have, and that there seems actually nothing, short of wartime destruction, which will prevent us getting all we need. But we need to use electricity wisely and carefully, and not take it for granted. We need to look to the manner in which we are able to convey it to the devices we want it to operate. We don't want to waste it, nor do anything which can interrupt its flow to or in our homes and offices. We want to use it more efficiently and practically, and that is what this book is all about—getting more service and pleasure from your magic electric genie. He is a giant in capability, but he does have a few rules and requirements which we must meet if he is to serve us properly.

In this book you will find lots of ideas, not only on electrical wiring but also on illumination techniques. Some may be new to you, some old. We hope that if you adopt just one of these ideas you may find that you have made a considerable savings, or prevented costly accidents or problems. To paraphrase an old saying: If just one of these ideas saves life or money or increases your pleasure in living, then the cost of this book will have been repaid a thousand times over.

This is not a book for just the man of the house. The lady and older boys and girls can gain knowledge and ideas from it. One word of caution, however. If you are **sure**—then do it. If there is **doubt**—call an electrician or electrical contractor! While our genie is friendly, he doesn't know his own potential. If his rules are ignored, misunderstood, or modified, he may without warning become a giant of destruction. Use care and he'll always be at your service, friendly, smiling, helpful. Learn the rules and follow them exactly. That's all he asks.

E.L. Safford, Jr.

ACKNOWLEDGMENTS

The following firms provided much pertinent information, including pictures used in this book. Although use of their product information herein does not reflect endorsement by the author or publisher, nevertheless, we express our gratitude to:

Arrow-Hart, Inc.
103 Hawthorn St.
Hartford, CT 06106
Tel: (203) 249-8471

Emerson Electric Mfg. Co.
8100 Florissant Avenue
St. Louis, MO. 63136

General Electric Co.
Wiring Device Business Dept.
95 Hathaway St.
Providence, RI 02904
Tel: (401) 781-1800

General Electric Co.
Large Lamp Dept.
Lighting Institute
Nela Park
Cleveland, OH 44112

Loran, Inc.
1705 E. Colton Avenue
Redlands, CA 92373

Tel: (714) 794-2121

Owens-Corning Fiberglas Corp.
Home Building Products Division
Fiberglas Tower
Toledo, OH 43659

Tel: (419) 259-3000

Wiremold Co.
West Hartford, CT 06110

Tel: (203) 233-6251

CONTENTS

Understanding the
Electrical System

The many differences found in electrical wiring systems usually reflect the date of installation. Older homes and offices are wired according to the standards and practices of the time, the type of construction, and funds allocated to the project. The electrical system generally represents about 3 to 5 percent of the total expense of a home or office and, for the most part, is designed to be "just adequate" for the planned use of the occupant. When we say "just adequate," this means, of course, that the planned circuit loading will be, or has been in proportion to the **code** which was in effect at the time the blueprints were made and approved.

CODES

The code is the governing factor which influences the wiring in your home or office. It is an established set of rules and procedures which were first formulated to insure that electricity would be a useful servant and not a wild killer. The primary purpose of the code's rules is to prevent fire which can so easily occur from faulty electrical installations, or death due to accidental contact with live wires. We shall discuss some pertinent code rules in the next chapter.

So what we mean by saying that the wiring will have been in accord with the code rules is that the **demand factor** will have been met. What is the demand factor? This is a number, the number of amperes (which is a measure of the flow of electricity like gallons per minute is the measure of the flow of gasoline) which you will use in your home under normal conditions. For example, you wouldn't plan to have all the range burners, the washer, the iron, etc., all on at the same time. If you did have everything that could run on electricity plugged into sockets or turned on, then you would have 100 percent demand of the flow of electricity for your home or office.

The wiring in your home will normally have been planned for some percentage of this total demand. And when the wiring

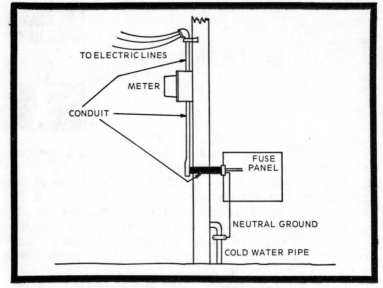

Fig. 1-1. An above-ground service installation.

is designed, the **size** of the wires used, the **kind** of wires used (normally **copper**) and the **number** of wires in a pipe, or nonmetallic sheathed cable, will determine how many amperes can be drawn from each circuit or outlet.

FEEDERS

Outside your home or office there are **feeder wires** (Figs. 1-1 and 1-2) that bring electricity from the pole through the meter(s) and appropriate fuses or circuit breakers into **your** home or office electrical system. The diameter of these feeder wires determines the amount of current that can flow inside the house.

SOME DEFINITIONS

At this stage we must become just a little technical. There are some terms you must be familiar with, if you are to understand and be able to talk about electricity. There are three **major** terms, and these you have probably heard before.

The Volt

This is the unit of **pressure** in the electrical system. It is abbreviated V. To have 10V is like having 10 pounds of water pressure. The more volts you have, the more electricity will be forced to flow, and the more **dangerous** the system becomes. Normally you will have 120V for lighting and at the receptacle outlets. You will have 220V for the range, the clothes dryer, and shop tools of large size like a lathe. The number of **volts** which you use is important when you ask for wire to add to your home circuits, or install new circuits in a new home. As you can readily imagine, if you use a wire which has a covering (insulation) designed to contain 10V, this same covering might rupture if you apply 120V pressure to it. Always be sure to buy wire which has a UL label showing it is tested for the voltage you plan to use it on.

An analogy might be useful if you are completely unfamiliar with electricity. If you have a water pipe and the wall of the pipe is too thin to withstand the internal water pressure, the pipe will rupture. With electricity, the **insulation** covering the wire is like the wall of the pipe. If the insulation has insufficient strength for the voltage pressure applied, then electrical short circuits, groundings, sparks, heating, blown fuses, etc. may result. Handling such a live wire may permit the electricity to leap from the wire into your body; and you must always be very careful when required to handle a live electrical wire! It may kill you!

Remember, when you buy wire for your home or office electrical circuits, be sure it has the proper volts value. It is

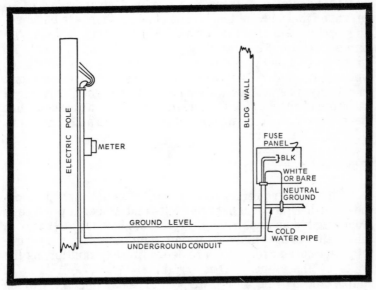

Fig. 1-2. An underground service installation.

customary to use wire which is insulated for 600V for all home wiring and office wiring. Ask your dealer about this.

The Ampere

This is the **flow** term. It is abbreviated A. It represents how much electricity is used each second. Here again we can imagine a pipe carrying water. If we want a fast flow, that is, a large number of gallons per minute through the pipe, then we use a large diameter pipe. With electricity, if we want a large flow, we use a large diameter wire. Any time we do not provide sufficient diameter pathway for the flow of electricity, the wire will get hot. It may burn and cause a fire if proper precautions are not used in installation. Always use the right diameter wire for the circuit.

Wire Sizes

Electrical wires are specified in terms of how many amperes can pass through them without overheating. For 15 to 20A circuits, such as are in your home, wire sizes are No. 12 and 14. For larger currents, sizes 10 and 8 are sometimes used. Notice that the wire size is larger as the number becomes smaller. For example, a wire size of 1 is much larger than a wire size of 22. In your electrical extension cords, the wire size is generally about 18 and this size can safely pass a current of 10A. Incidentally, if there is a group of stranded wires so arranged that they all touch each other, so they are bunched, then their size is equivalent to a solid wire of copper having the same diameter. The stranding just gives flexibility. Table 1-1 gives the current-carrying capacity for common wire sizes you will encounter. **Bell wire**, which is used around the home, is usually size 20, but its **insulation** is designed only for a pressure of about 20 – 24V. So it will work fine on doorbells, but never, never use it for any circuit connected to a 120V outlet in your home or office.

The Watt

This is the **power** term. It is abbreviated W. It represents how much **equivalent** heat the volts and amperes produce if connected to an appliance. It represents how much energy is dissipated as light and heat from a light bulb. It is a measure of power consumed. Watts are determined by multiplying the volts by the amperes. The formula is $W = V \times A$. This is important to us because if we plan to use an appliance which has a rating of 240W at 120V, then we can find the amperes

Table 1-1. Wire Size vs Ampere Capacity

WIRE SIZE NO.*	AMPERES
18	10
16	15
14	20
12	30
10	40
8	55
6	80
4	105
2	140
0	195
00	225
000	300

Note that the capacity increases as the wire number decreases. The above ratings are for **copper wire**, either solid or stranded. **We do not recommend aluminum wire in any installation**.

* American Wire Gage (AWG)

which the circuit will have to be able to pass. Since the original formula is W=V x A, we must transpose to find A. By algebraic manipulation, the formula becomes A=W/V. We then plug in the numbers and have

$$\frac{240W}{120V} = 2A.$$

By the way, 240W is about right for a small tube-type TV set. Since most home outlets individually are connected to pass at least 15A, if nothing else is on the circuit, we can use the TV and other appliances or lights on the circuit without problems. If you use an iron to iron clothes and it is rated at 1000W, then note that the current is high:

$$\frac{1000W}{120V} = 8.33A.$$

This is about the most current you should expect to pass through an 18-gage cord without excessive heating and its resultant power waste.

Watts is the **rate of use** of electric power. Here is a definition to explain watts in terms of the amount of work they can do for your. One horsepower is equal to 550W. One horsepower is equal to 33,000 ft lbs. This means the energy it takes to raise a load of 33,000 lb a height of 1 ft. One horsepower is also

13

the effort of one strong, fast horse pulling 330 lb up out of a well that's 100 ft deep in 1 minute!

Summary

Volts is the important term meaning pressure that forces current to flow and determines the amount of nonconductive covering of the wires—the insulation. We must always use wire with sufficient insulation to prevent breakdown.

Amperes is the flow of electricity. If too much flows in too small a wire, the wire gets hot, can burn the insulation, cause a breakdown (short circuit) and blow fuses or cause fire.

Watts relates to power used. It is a term noted on electrical appliances and devices to tell us what kind of circuit we must have to operate the device satisfactorily. For example, a 100W, 120V light bulb will produce 100W of heat and light with a current of 0.83A from a 120V line. Table 1-2 shows some wattage values for various appliances and Fig. 1-3 shows a modern kitchen with its normal appliances. Fig. 1-4 shows a modern family room lighting arrangement.

Table 1-2. Some Wattage Values for Various Devices

APPLIANCE OR DEVICE	WATTS REQUIRED	APPLIED VOLTS
100 watt light bulb	100	120
Electric range	12,207	220
Clothes dryer	4,856	120 (220)
Window air conditioner	1,566	120
Broiler	1,436	120
Dishwasher	1,201	120
Electric frying pan	1,196	120
Laundry iron (home type)	1,088	120
Vacuum cleaner	630	120
Blender	250	120
Automatic washing machine	512	120
Furnace which is fuel fired	800	120
Color television	332	120
Sewing machine	75	120
Radio (depends on size, normally)	70	120
Garbage disposal	900	120
Hot water heater	2,500	220
Refrigerator	250	120
Coffee maker	600	120
Toaster	1,100	120

Note that although a 100W bulb at 120V consumes only about 1A (0.83A) the same wattage bulb operated from your 12V car battery would require **ten times the current** or 8.3A. Thus the wire required to satisfactorily operate the same wattage bulb on the lower voltage system would have to be at least No. 18. Low voltage systems are used in outdoor lighting systems for homes and office buildings because the pressure (volts) is less and the danger thus reduced.

CIRCUIT OVERLOAD

What happens when you overload an electrical circuit? Blow a fuse? Exactly! That is because each circuit is protected. What then is a "circuit"? Look at Figs. 1-5 and 1-6, and Table 1-3.

When we overload the circuit by placing too many lights or appliances on the "line," the wires will run hot. Yes, actually they will get physically hot just like the heating element in your electric range, toaster, space or water heater. This heat overload can cause the insulation to melt inside the conduit, then permit wires to touch each other and cause short circuits. A short circuit is like opening the floodgates to a dam—all electricity possible will flow, increasing the temperature, creating tremendous heat which may even heat the

Fig. 1-3. A modern, well lighted kitchen with all appliances. (G.E.)

Fig. 1-4. A modern, well lighted family room. Note the absence of glare. (G.E.)

conduit to such a level that something placed against it in the home construction may catch on fire. WE NEVER WANT TO OVERLOAD CIRCUITS EVEN THOUGH THEY ARE FUSED.

What does all this mean? It means that if your home is an older one **and you keep adding new appliances**, lights, etc., then you need to consider new circuits, additional circuits, larger feeders, etc., so that you will have a safe and effective installation. You will also have better operation from your television and all other appliances, and better light, for you

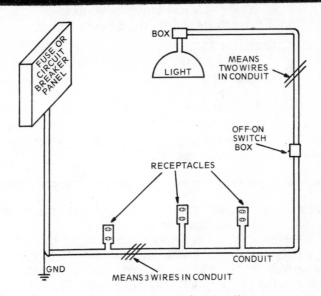

Fig. 1-5. The concept of a circuit.

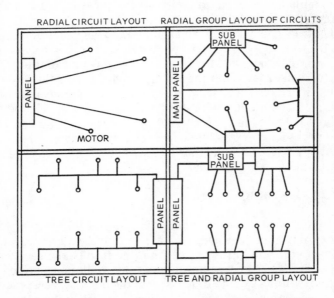

Fig. 1-6. Types of circuits used in modern home and office installations.

**Table 1-3. Load Limits in Watts for
Various Capacity Circuits**

CIRCUIT CAPACITY (AMPERES)	LOAD LIMIT (WATTS)
15	1800
20	2400
30	3600

Note: You can add the wattage of the lights and appliances on a circuit to determine total circuit loading.

will have provided sufficient capacity so that the current can flow to all devices without loss.

When we overload circuits we can place such a demand on them that not every device can get enough electricity to operate as it should. The result? Poor television performance, faulty appliance operation, and dim, flickering lights. All of these may be present at the same time when there is insufficient capacity in the wiring, or the power company has not supplied sufficient capacity on the pole lines and transformers for the neighborhood. One good fact—you can always call the electric company and ask them to run a check on your electricity supply. They will put a graph recorder on the line outside your home and let this run for a day or two, then they can tell you exactly how the voltage varies during the day and night (if it does) and what the amount of voltage is at any instant. Needless to say that if they run such a check and the voltage is low—they will correct the problem. They can also run checks for electrical interference which may be caused by old, corroded, etc. connections on the power poles. If you have this trouble, call them also.

CIRCUIT SYMBOLS

We will now discuss briefly how to find circuits in your home. A circuit consists of all the lights and receptacles connected to one pair of wires from the fuse box. These were shown in Figs. 1-5 and 1-6. On paper, a circuit is described by certain symbology which shows what the devices, receptacles, switches, lights, etc. are, and where they should be located. Refer to Table 1-4 for this type of symbology.

Notice how wires are designated when there are more than one. Find the single receptacle, duplex (most common)

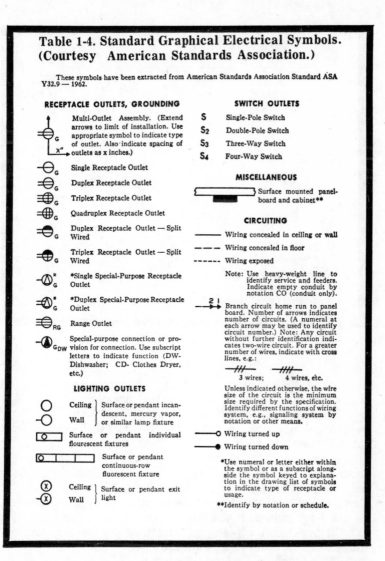

**Table 1-4. Standard Graphical Electrical Symbols.
(Courtesy American Standards Association.)**

These symbols have been extracted from American Standards Association Standard ASA Y32.9 — 1962.

RECEPTACLE OUTLETS, GROUNDING

Multi-Outlet Assembly. (Extend arrows to limit of installation. Use appropriate symbol to indicate type of outlet. Also indicate spacing of outlets as x inches.)

Single Receptacle Outlet

Duplex Receptacle Outlet

Triplex Receptacle Outlet

Quadruplex Receptacle Outlet

Duplex Receptacle Outlet — Split Wired

Triplex Receptacle Outlet — Split Wired

*Single Special-Purpose Receptacle Outlet

*Duplex Special-Purpose Receptacle Outlet

Range Outlet

Special-purpose connection or provision for connection. Use subscript letters to indicate function (DW-Dishwasher; CD- Clothes Dryer, etc.)

LIGHTING OUTLETS

Ceiling } Surface or pendant incandescent, mercury vapor, or similar lamp fixture

Wall

Surface or pendant individual flourescent fixtures

Surface or pendant continuous-row fluorescent fixture

Ceiling } Surface or pendant exit light

Wall

SWITCH OUTLETS

S   Single-Pole Switch

S_2 Double-Pole Switch

S_3 Three-Way Switch

S_4 Four-Way Switch

MISCELLANEOUS

Surface mounted panelboard and cabinet**

CIRCUITING

——— Wiring concealed in ceiling or wall

– – – Wiring concealed in floor

------ Wiring exposed

Note: Use heavy-weight line to identify service and feeders. Indicate empty conduit by notation CO (conduit only).

2 1
Branch circuit home run to panel board. Number of arrows indicates number of circuits. (A numeral at each arrow may be used to identify circuit number.) Note: Any circuit without further identification indicates two-wire circuit. For a greater number of wires, indicate with cross lines, e.g.:

///   ////
3 wires;   4 wires, etc.

Unless indicated otherwise, the wire size of the circuit is the minimum size required by the specification. Identify different functions of wiring system, e.g., signaling system by notation or other means.

—O Wiring turned up

—● Wiring turned down

*Use numeral or letter either within the symbol or as a subscript alongside the symbol keyed to explanation in the drawing list of symbols to indicate type of receptacle or usage.

**Identify by notation or schedule.

receptacle, ceiling light, and switch symbols. See Fig. 1-7. You might note that a **single pole switch** is simply an off-on switch, which opens one wire or connects it together. A double pole switch may open two wires simultaneously and connect them back to their interrupted lengths.

LOCATING INDIVIDUAL CIRCUITS

Now, how can you locate the circuits? One way to do this is as follows:

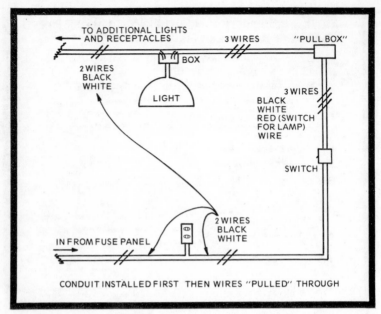

Fig. 1-7. Concept of installation of conduit, rigid or flexible. The conduit is installed first, and then the wires are pulled through it.

a. First get a good flashlight. Then turn off **all** circuit breakers or remove **all** fuses in the fuse box but one.

b. Go around the house turning all lights on. Only those which are connected to this circuit breaker will light. Make a note of them, i.e., Hall, Hall Bath, Bedroom, etc.

c. Take a table lamp and turn it on. Now go through the house plugging it into each outlet receptacle. Only those connected to the same circuit as the lights in (b) will cause the lamp to light. Make a note of the location of these receptacles.

d. Now go back to the circuit breaker box. Turn this first circuit breaker (or fuse) off and turn the second one on. Repeat the procedure as stated above.

Once you have located the circuits, then you know which fuse to loosen or which circuit breaker to turn off to work on a given circuit. You may want to modernize to improve the lighting. (See **before** and **after** photos of Figs. 1-8 and 1-9.)

TYPICAL WIRING LAYOUTS

Sometimes the wiring is run as in Fig. 1-10, the electrical layout of an average home. In some cases, the wiring is run overhead in the attic in flexible or rigid conduit, or may be two lines which are separated by about 12 in. Connections to these lines will come down between the joists to the location of the switches and receptacles. An example is shown in Fig. 1-11.

In new home wiring, flexible conduit is a common method of electrical installation. When all the BX (flexible metal conduit) is in place and connected to all outlet boxes, then the electrical wires themselves are pulled through the conduit, connected to devices at the boxes, and the electrical installation is complete. Rigid conduit may be used, or a flexible two-conductor sheathed cable which is nonmetallic. One trade name for this nonmetallic cable is Romex.

On Fig. 1-10, locate the S3 switches. These are two-way switches which permit the light over the stairs to be turned on and off from either upstairs or downstairs. Much modernizing is done by installing these two switches and a light either on stairs, in halls, between the entrance to a garage and the

Fig. 1-8. A poor lighting situation. (G.E.)

Fig. 1-9. With some additional lighting, there are no dark shadows. (G.E.)

house, or from the yard to the house. The idea is that you can turn the light on (or off) from **either** position. Three-way switches are in the kitchen, living room, garage, etc.

Note the arrangement of outlet receptacles. They should be placed around the room so that there is no position along the walls that will be over 6 ft away from an outlet. The reason for this is that most electrical appliances have 6 ft cords and in a properly designed home or office, you should not have to use extension cords.

Pull-chain lamps are sometimes used in halls or closets for economy reasons. There must **not** be a pull chain, where the chain is metal all its length, **in a bathroom**. It's best in the bathroom to use wall boxes and switches for everything. Electric heaters placed in a bathroom may invite electrocution. They should be permanently mounted into the wall and the frame grounded to the water system for maximum safety. Use of electric hair dryers, etc. in the bath or kitchen where one may touch a water pipe may result in injury. Use these devices in the **bedrooms**, or be sure they are grounding types. These types have a **3-prong plug**. (Chapter 2 will provide a full description of **grounded** and **grounding**.)

Now let us examine Fig. 1-12. This could be the most modern type of home wiring, which permits the turning on of room lights from every doorway, etc. The symbology for Fig. 1-12 is listed in Fig. 1-13, so you can understand what kinds of switches and remote devices are used. In Chapter 9 you will find more details on the kind of low voltage switching and the devices used in this particular kind installation.

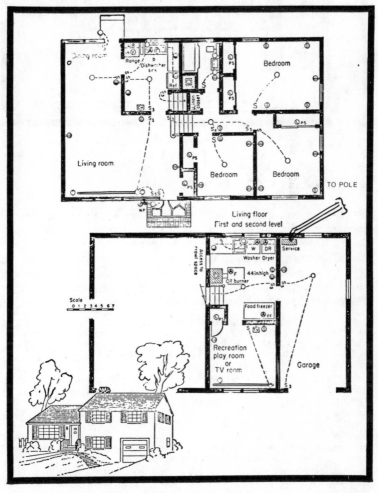

Fig. 1-10. Architect's plan for home wiring.

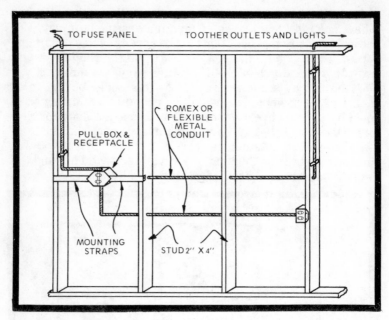

TO FUSE PANEL   TO OTHER OUTLETS AND LIGHTS →

ROMEX OR
FLEXIBLE
METAL
CONDUIT

PULL BOX &
RECEPTACLE

MOUNTING
STRAPS

STUD 2″ X 4″

Fig. 1-11. Method of running flexible metallic conduit in the wood framing of rooms. Sometimes the conduit may be run in "notches" on the room side of the studs (joists).

COMMON CIRCUITS IN THE HOME

Let us consider some of the modern wiring practices to gain an insight into just how circuits are divided in your home. Items like the water heater, the furnace, the overhead air conditioner (evaporative or refrigerated) will probably be on separate, individual circuits. Each heavy duty item may be operated from a separate 220V line, depending on total loads. The electric range also may have its own circuit. Normally the

Fig. 1-12. A deluxe wiring system employing the use of multipoint switching that permits turning on room lights from every doorway, switch-controlled split receptacles in living room and master bedroom, master-selector switch at the bedside. Selected lights and receptacles are also controlled from two motor-master units not shown. It has been designed to provide protection control, turning on perimeter lights to floodlight the grounds, as well as lighting selected rooms from the master bedside. Pressing a single button(Sm)at any entrance, or the 12th position of the master-selector switch starts the motor-master for **on** control of the selected lights to provide a pathway of light through the house, or the **off** side for turning off all the lights in and around the house. And here's downright convenience: a switch-controlled outdoor weatherproof split receptacle that lets the homeowner turn the decorative Christmas lights on and off from two handy indoor locations. Any portion of this layout can be used as a guide in laying out a G.E. remote control installation to meet requirements specified by the architect, builder, or home-buyer.

24

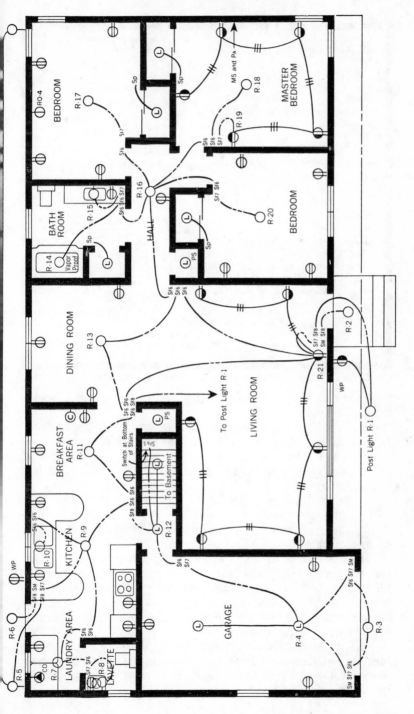

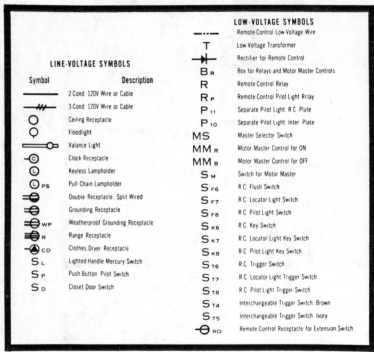

LOW-VOLTAGE SYMBOLS

Symbol	Description
----·---	Remote Control Low Voltage Wire
T	Low Voltage Transformer
⟶▸⟵	Rectifier for Remote Control
B R	Box for Relays and Motor Master Controls
R	Remote Control Relay
R P	Remote Control Pilot Light Relay
P 11	Separate Pilot Light, R C Plate
P 10	Separate Pilot Light, Inter Plate
MS	Master Selector Switch
MM R	Motor Master Control for ON
MM B	Motor Master Control for OFF
S M	Switch for Motor Master
S F6	R C Flush Switch
S F7	R C Locator Light Switch
S F8	R C Pilot Light Switch
S K6	R C Key Switch
S K7	R C Locator Light Key Switch
S K8	R C Pilot Light Key Switch
S T6	R C Trigger Switch
S T7	R C Locator Light Trigger Switch
S T8	R C Pilot Light Trigger Switch
S T4	Interchangeable Trigger Switch Brown
S T5	Interchangeable Trigger Switch Ivory
⊖ RO	Remote Control Receptacle for Extension Switch

LINE-VOLTAGE SYMBOLS

Symbol	Description
⸻	2 Cond 120V Wire or Cable
⸻	3 Cond 120V Wire or Cable
◯	Ceiling Receptacle
◯	Floodlight
⬤⟳	Valance Light
◉C	Clock Receptacle
◎	Keyless Lampholder
◎ PS	Pull Chain Lampholder
⟩◉	Double Receptacle, Split Wired
⟩◉	Grounding Receptacle
⟩◉ WP	Weatherproof Grounding Receptacle
⟩◉ R	Range Receptacle
◉ CD	Clothes Dryer Receptacle
S L	Lighted Handle Mercury Switch
S P	Push Button Pilot Switch
S D	Closet Door Switch

Fig. 1-13. The symbology used in Fig. 1-12.

fuse box or circuit breaker wiring identification will tell you this information.

But as to the other circuits, and remember they are wired for symmetry, they are probably grouped as follows:

a. When there are small bedrooms close together with a bath close by or in between, these will be on one circuit for lights and receptacles.

b. The master bedroom and its bath and perhaps a hall will be on one circuit.

c. The living room and possibly the kitchen lighting can be on one circuit.

d. When the dining room and the kitchen are close together, you may find the outlets on a common circuit, although remember there must be a 20A circuit individually in each of these two rooms. The dining room lights may be on the same circuit as the garage if these are close, or the hall lights may be on the dining room circuit.

e. Basement circuits are usually individually run because there are often workshop devices connected there. The lights, however, may be grouped with another circuit for rooms close by, on the ground floor.

f. Dishwasher, garbage disposal, laundry washers and dryers all may be on separate circuits or, depending upon the demand in current, may be grouped in only two, such as the dishwasher and garage disposal together.

In general, the plan is to distribute the circuits so that there will not be over the 12 to 20A demand on living room, bedrooms, etc., and the 20 to 30A demand on the basement or workshop. All other large electric items have individual circuits.

There is another consideration which is important from a user viewpoint. That is, the knowledge that all outlets in bedrooms, living rooms, etc. are wired so that at least 15A can be drawn from them **if there is no other drain on the circuit**. You will recall that 15A at 120A is 1800W, by simple multiplication. You will also realize that the normal clothes iron draws about 1000W. So in terms of devices, the 1800W would be two of these irons operating on the same circuit. The idea here, then, is to be careful when you have a heavy-drain electric appliance. Use it on a very lightly loaded circuit.

How are the wires run from the fuse or circuit breaker panel to the various circuits? This is done by having a base point in each circuit, called the **center of symmetry**, usually, to which the primary pair of wires from the fuse panel goes. This base point will be a box such as the overhead light receptacle in a bedroom. All other branch circuit lines for this circuit will then run to this common point and connect to the main pair of wires.

The run from the center base point to the fuse box is called the **home run**. A home run will be two wires, one white and one black, of large enough size, No. 14 to 10 usually, to handle all the load on the circuit. Refer to Fig. 1-6 for the example circuitry.

There can be as many as 25 home runs in a single home. Offices will have just as many, if not more, depending upon the kind of lighting and the kinds of electrical equipment to be found there.

When an architect sends his home plans to a consulting electrical engineer for the wiring and lighting planning and details, the layout which was shown in Fig. 1-10 results. The sizes of wires will be specified, the location of panels and switches, the kinds of lights, etc., will all be included. Next the plan goes to the contractor who has several electricians bid on the work. In this step every nail and screw, every length of conduit, every switch, etc. are all added to the **time** it takes to make the installation and the result is the cost to you, the home builder, of the total electrical installation.

Table 1-5. Number of Conductors Permitted in Various Sizes of Conduit

CONDUCTOR WIRE SIZE	CONDUIT DIAMETER, INCHES					
	½	¾	1	1-¼	1-½	2
18	7	12	20	35	49	80
16	6	10	17	30	41	68
* 14	4	6	10	18	25	41
* 12	3	5	8	15	21	34
* 10	1	4	7	3	17	29

Note: Sizes 14, 12, and 10 are those normally used in home circuit wiring.
The conduit size in homes varies from ½ to 1 ¼ in. usually.

Of course you know this. The reason we mention it is that the **size** of the conduit, the number of wires in a circuit, the number of home runs are all based on the electrical receptacles, lights, etc. specified by the consulting engineers and the architects. This means that normally you **cannot run more wires** through an installed conduit to add new wiring! **If you add MAJOR circuits you normally have to make a new home run and then add the additional outlets to the new circuit.** It is possible, of course, that if you just want a few outlets added, and these would not measurably increase the load on an existing home run circuit, but would simply be convenience type outlets or lights which would be used **in place of**, and not in addition to, the others in the home, you CAN add to the existing outlets as we will show in Chapter 5. But be aware of the precautions. If you are **increasing the demand on a circuit, then you need a new home run to the fuse box**, which means a **new wiring circuit** in the house.

We examine the number of wires permitted under the code (to be discussed in the next chapter) for various size conduits—either flexible-metal, rigid-pipe, or plastic types. Thus, we can look into an outlet box, see how many wires are there, see what size the conduit is, and find out if we COULD run another wire in this conduit without exceeding the code limitations. Table 1-5 shows the relationships.

The reason for the limitation on number of wires is heating. If we run enough wires in a pipe so they form a tight fit, and then use the maximum current these wires can carry, heat will build up from each wire inside the pipe and could cause the insulation to burn. We'd likely have smoke and blown fuses and perhaps have to install new wiring. So it is really cheaper in the long run to install the new wiring, if needed, rather than to have to remove the old wiring out of damaged pipes or conduit—then run new wires, etc. The table will tell you what you can do, whether you can add new wire, or have to install new circuits. Be guided by it.

SAFETY

Now, since we've opened the door on safety, we'll consider some ideas which should never be forgotten. We are talking now about the circuits which we obtain through the use of extension cords. These can be most dangerous, and cause disasters, which can occur because we human beings are apt to be forgetful. Frequently, it is misuse of electrical equipment that causes shocks, burns, and fires. For example, extension cords should not be run through holes in walls or floors, strung through doorways or windows or put under a rug, where constant wear could rub off the insulation and start a fire.

Many a fire starts because someone forgot to unplug the iron, and many a person gets shocked because he forgot to unplug the toaster or other appliance before cleaning it. And remember, it is extremely dangerous to use any electrical equipment while standing on a wet surface or even while perspiring heavily.

Finally, to wind up this chapter, we want to present some solutions to some common electrical problems. These all might be grouped under a heading of "what to do when it won't work." The procedure goes like this:

1. Make sure you've turned the switch on, that the thermostat (if used) is set at an operative temperature, and that the equipment is plugged in.

2. If that's not the problem, unplug the disabled equipment and, if it's a 120V appliance, plug a lamp into the same outlet to see whether the outlet is working. If it isn't, check your fuse or circuit breaker box. **Standing on a dry board or mat,** turn off the house current and replace any burned out fuses or reset the circuit breaker(s). If restarting the appliance blows a

fuse or trips a circuit breaker again, it could mean you've overloaded the circuit. To test for an overload, unplug other appliances on the same circuit.

Sometimes overloads occur during the brief period when a motor starts up and requires an extra surge of power. In that case a **time-lag fuse with the same ampere rating** can solve the problem.

Never replace a fuse with one of higher amperage!

3. If the outlet is live when you check it, the trouble probably is in the appliance. Be sure to keep it unplugged while you examine or repair it. Check the plug prongs; if they don't hold securely in the outlet, bend them slightly outward. Check the cord to be sure it is securely fastened to the plug and appliance terminals. Examine the cord, switch, and plug for wire breaks or damaged insulation.

4. **Reset buttons and emergency switches** on furnaces, air conditioners, garbage disposals, water heaters, electric motors—pushing in a reset button or turning on an emergency switch that inadvertently has been moved to off is frequently all it takes to get an appliance running again. Be sure the button is fully depressed.

5. **Thermostat controls**—make sure they are not set too low or too high for proper operation and that they aren't located where they can be thrown out of kilter by drafts, abnormal sources of heat, or objects bumping against them.

6. **Fuses in an appliance**—if the convenience outlet or timer assembly on an electric range fails, check the small fuse, often beneath one of the rear burner units. On some older dryers, there may be a small fuse behind the back panel to be checked if the appliance won't run.

7. **Bent blades and loose bolts on fans or food mixers**—straightening blades may solve the problem of a motor that overheats or strains when running. Tightening bolts can eliminate rattles, vibrations, and laboring of motors.

Codes, Rules, and Regulations

The **Code** is the list of electrical standards which govern electrical installations. Such standards have been developed and are sold by the National Fire Protective Association, Batterymarch Street, Boston, Mass. 02110. They do not carry the weight of law, but they do form the basis for almost all of the city ordinances and inspections and regulations. A copy of the code book costs but a few dollars and is well worth the expenditure. Or ask your librarian for the National Electrical Code (NEC).

The primary purpose of the code is to protect life and property by specifying the best and safest way to install electrical systems. As we examine some parts of the code we will gain more insight as to the kind of wiring permitted, the number of wires to use in a conduit, the size wire which must be used for given loads, etc. Since we have had some concept of this in the first chapter, we go into a little more detail here.

FIGURING THE LOADS

The **load** requirement in a home or office is the basic governing factor used to determine the size of wire, the number of wires in a given size conduit, the size and number of circuit breakers or fuses, and the size of the service input and panel. The load is always specified in terms of **watts required**, and this, in turn, is broken into the voltage necessary, and the amperes which will flow to produce a given number of watts at that voltage. We need to have some general guidelines regarding the number of watts **required** for lighting, and for other purposes.

When we consider general illumination in a home and office, there is a general rule which applies. In homes you will need 3–4W (watts) of light per square foot. In an office you will need at least 5W per square foot. To determine the area in square feet to be considered, you compute by measuring the **outside** dimensions of the building, apartment, or whatever structure you are concerned with. Open areas such as porches, garages (unless a part of the dwelling) or other unfinished spaces ordinarily are not included in the measurement. Once you know the outside dimensions, you simply multiply the length by the width to get the square feet of floor space. The manner in which to calculate the number of branch circuits to be used **in the structure** is then as follows: First, we can calculate the number of square feet and multiply this by 3 (watts per square foot). Thus, for example, if we found the structure had 2500 square feet—and this would be a rather large house, probably with four bedrooms, living room, family room, a couple of baths, and a garage attached—the illumination **minimum** requirement would be 2500 x 3 = 7500W. Since we would use 120V lines, this would mean

$$\frac{7500W}{120V} = 63A$$

of current (rounding **up** from ½). Since we would use 20A circuits, this would mean that we would require **at least four circuits for lights alone.** We must add the wattage from appliance receptacles to this. Also, for future expansion, it might be better to use 2500 x 4=10,000W divided by 120 equals 84A.

So, using the 3W example. as we plan a system of **outlets** for lights, radios, etc., **and light fixtures,** we would plan to consider 1.5A for each convenience outlet on a single circuit, and add the wattage of lights. Notice that each outlet is then considered to furnish power of 1.5 x 120 =180W. We know that the total **power** permissible on each circuit is

$$\frac{7500W}{4} = 1870W.$$

If we plan one circuit of a permanent nature using lights totaling 1000W (four 250W bulbs) then we could plan for only 4 receptacle outlets (4 x 180=720W) on this circuit. By the way, we **must** plan one receptacle for each 6 **linear** feet of wall space. Receptacles should never be over 6 linear feet away from one another for proper home convenience. This may mean more circuits are required than the four we calculated for lighting alone.

SPECIAL OUTLETS

When there are large appliance loads or heavy duty lamps or other devices such as motors to be connected to outlets, then the rule is to calculate the outlet amperage to be that value required for the appliance, and run separate circuits for each of them, if necessary. Sometimes a heavy duty outlet is installed and this simply means the circuit is designed so 5A (5 x 120=600W) is available here always. Four such outlets on a 20A circuit is all the load that is permitted. As we shall see later, there are requirements in a home that at least one outlet in kitchen and dining area have a full 20A capacity. This means only this outlet on one circuit. The amperage, of course, determines the size of wires to be used in the installation. Number 14 wire is the minimum size used, while No. 12 is better and No. 10 is used for heavy loads. Please refer back to the tables in Chapter 1.

GROUND AND GROUNDING WIRES

Let us discuss the actual wiring of a circuit. The code says that it is necessary to have a common "ground" wire in each circuit, and that these will all be connected together without interruption and connected to the service feed common wire, or neutral, or ground. Even though metallic conduit may be used throughout the system, this conduit itself CANNOT be used for the ground. The ground wire concept is shown in Fig. 2-1.

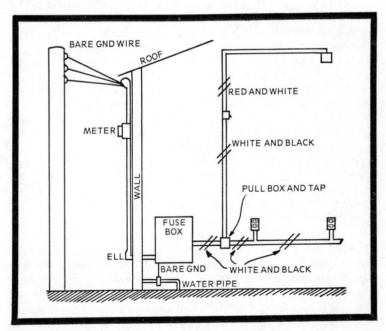

Fig. 2-1. The wiring code used in home and office wiring. Never put a switch in the **white** wire.

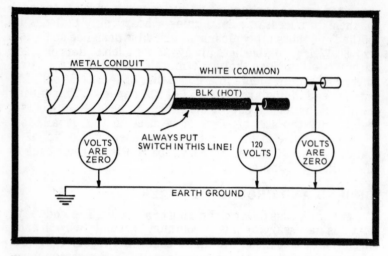

Fig. 2-2. If a switch is placed in the "common" or white line, then the circuit is always hot. Touching a circuit wire beyond the switch, a black wire with insulation off of it, can create a current to earth ground through your body, and could kill you.

The code says that the identity of the ground wire must be preserved and this is done by using a specific color for this wire. The color is **white** or a **gray** color very close to white. This color will **not** be used for any hot leg, or switch tap, or any other purpose. It is the COMMON electrical conductor in the circuit. The other wire in a circuit (this is the hot wire), which **can** be broken by a switch, etc., is **black** in the run to the fuse box, and **red** after it is broken by a switch. If we placed a switch in the ground wire, a dangerous situation could result (Fig. 2-2), so this must **never** be done.

In some older homes, which have been wired with two conductors which are possibly discolored by time, and are separated about 12 in. apart, and fastened to porcelain knobs and run through porcelain tubes, it is necessary to identify the ground wire so you may plan an extension or tap to, say, operate another light through a switch. Also, sometimes, in your home or office, an extension may have been made to existing wiring, or the initial wiring done by someone who either ignored the code or was unaware of it, and most any colors might be used in the circuits. (It has been my personal experience to find **two black wires** in a circuit run instead of one black and one white wire.) In each case, you can find out which wire is the ground wire by testing the circuit as shown in Fig. 2-3. BE CAREFUL and be sure to have adequate insulation on the probes.

There has been some question regarding the meaning of grounded, and grounding wires. It is important enough that we consider it here. The **grounded wire** is the common (white) wire throughout the system which is connected to a water pipe or other form of earth ground, and to the common bare wire of the feeder input. This wire does pass current in systems. The **grounding wire** is a green wire or a bare wire in **Romex** or other cable. You will find that it is the free wire on appliance cords or shop equipment cords. It connects the frame of metal equipment to the conduit or the common **grounded** wire in the system. **It carries no current** (normally) and is merely for protection. In case the frame of an electric appliance or device becomes hot to ground, a short circuit will take place—not through you, however, but through this grounding wire. Fuses will blow and thus tell you there is a **defect in the equipment.** If the grounding wire is not used (even though it should be), a short circuit that takes place

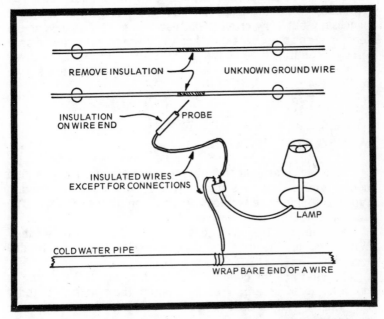

Fig. 2-3. Making a tester out of a regular small lamp. This can help you identify the ground (common) wire when the insulation does not have a proper color, or there is doubt as to which wire is the common one. BE CAREFUL!

  a. Turn the circuit off. Remove small sections of insulation.

  b. Connect one wire from a known good lamp to a water pipe. CAREFULLY touch the other wire from lamp to each bare place on unknown wires in turn. The lamp will light when the hot wire is touched. The circuit must, of course, be energized for this test.

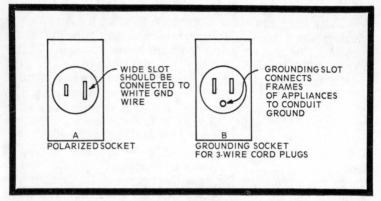

Fig. 2-4. Grounding type sockets. The first (a) is called a "polarized" socket; (b) is a grounding socket for 3-wire cords.

inside the equipment (making the frame hot or electrified while you are holding or touching it) could kill you! The **grounded wire is required.** The grounding **wire** may be necessary but very often is not used—**use it!** Check your extension cords and see if they are the 2-wire **nongrounded** or 3-wire **grounding** type. GET AND USE THE **GROUNDING TYPE!** A polarized socket and a grounding socket are shown in Fig. 2-4.

WIRING SYSTEM TYPES

Now we return our thoughts to the types of wiring used in homes and offices. We find that the code specifies certain types of wiring based on four general conditions of usage. The code considers the following situations:

a. What are the conditions of the location of the wiring? Will it be on the exterior of buildings? Will it be in dry locations?

b. What are the conditions of the location of the wiring in building construction? Will it be under the plaster, or under the plaster finish; will it be embedded in the masonry or concrete?

c. What are the intentions of occupancy of the building which is wired? Will it be a home, office, farm, what? Who and what will be contained in it?

d. What is the purpose of the wiring? Will it be to operate shop machinery, will it be normal home use? Normal office use? Will it be on a farm in a barn, for lighting, equipment operation? Will it be used in a hazardous location where gas fumes are present?

The various types of wiring systems currently in use are listed in Table 2-1. Most all types are found in homes and offices. Item 17 is used in hazardous locations and underwater or in damp locations. Item 15 is used for underground feeders or wiring extensions in the yard. Busways are commonly found in industrial locations or commercial buildings as are items 10, 19, and 20. In the home, the most common are the rigid or the flexible metal conduit, and the nonmetallic sheathed cable such as Romex. In some areas plastic tubing of a rigid type is approved as the container of the wiring to replace the metal conduit. There is a disadvantage with this. There is no continuous grounded cover for all wiring as can be obtained with metal conduit. You will have to run a separate grounding wire in the system using plastic tubing to which you ground the outlet boxes, provide connection for 3-prong grounding cords, etc. We also suggest that when you plan new wiring, or or changes to old wiring, you call your city inspector to see if

---

Table 2-1. Acceptable Types of Electrical Wiring

1. Busways—Metallic box-like enclosures which have heavy metal bars inside, along which currents flow. They are for very large current demand installations.
2. Wireways (raceways).
3. Underfloor raceway.
4. Surface metal raceway.
5. Flexible conduit (metal) BX.
6. Rigid conduit (pipe).
7. Nonmetallic surface extensions (Romex, etc.).
8. Electrical metal tubing.
9. Armor cable.
10. Bare conductor feeders.
11. Service knob and tube work.
12. Concealed knob and tube work.
13. Open wiring on insulators.
14. Mineral-insulated, metal-sheathed cable, Type MI.
15. Underground feeder and branch circuit cable Type UF.
16. Underplaster extensions.
17. Liquid-tight, flexible metal conduit.
18. Multioutlet assembly.
19. Cellular floor raceway.
20. Cellular concrete floor raceway.
21. Nonmetallic sheathed cable (Romex, etc.).

there is a particular type of wiring system which he requires for the change you have in mind. Remember you do this not only to comply with city ordinances, but also FOR YOUR OWN SAFETY! It is unwise to save a few dollars on wiring and have it cost you thousands through a fire in your home! And unapproved wiring that causes a fire may nullify your fire insurance!

The code says that under certain conditions, exposed metal parts of fixed electrical equipment which do not carry current but which are liable to become energized, **shall be grounded.** The word **shall** means that it is a requirement and mandatory. These conditions are:

1. Where equipment is supplied by means of metal clad wiring.

2. Where equipment is located in a wet location and is not isolated.

3. Where equipment is located within reach of a person who can make contact with any grounded surface or object. This is important in bathrooms, kitchens, and laundries. There one finds water pipes which are the BEST kind of grounds. Also there is water and dampness which help form conducting media. So **all fixtures** in these areas should have **metal covered grounded plates.** This means switch covers and receptacle covers must be metal and **not plastic.**

4. Where equipment is located within reach of any person standing on the ground.

5. Where equipment is in hazardous locations.

6. Where equipment is in contact with a metal building lath or other metal.

7. Where equipment operates in excess of 150V. This means the 220V outlets and switches.

8. Frames of motors and controllers must be grounded.

9. Electric equipment in garages must be grounded.

Generally speaking, a voltage of not over 220V is used in the systems of wiring discussed and listed here. We normally have just 120V in our homes and offices, except for ranges, etc., which use 220V. In some offices, equipment such as computers may use 440V. All the electricity we use is the **alternating kind.** This simply means that we can use **transformers** to increase or decrease its voltage value. The use of a bell ringing transformer is an example of voltage reduction from 120V to, say, 6 or 12V.

SAFETY REGULATIONS

We must **never** run any other kind of signaling or convenience circuit wires in the same conduit with electrical power wires. If we did, and IF somehow a short circuit took place between the power wires and these other kinds of wires, we might have an **electrocution** on our hands. One example that comes to mind might be the running of telephone cables in the same conduit with electrical power wires. This is **never done** by professionals. Also if you have seen what appears to be these two types of wires in the same conduit, please examine them again; you will find a sealed, **divided** conduit which keeps these different kinds of wires isolated completely one from another. The codes say that all wiring must be protected against physical damage, corrosion, or any physical happening which might expose the wires or electrify anything which could shock or kill someone. Conductors for **open wiring** must be rubber covered, thermoplastic, slow burning, weatherproof varnish cambric, insulated slow burning, or asbestos covered. These wires are run on knob insulators and through porcelain tubes when the wires run through wood frames, walls, etc.

The metal conduit and metal boxes of an installation must be mechanically and electrically **continuous** so that an electrical ground is made continuously through them. When locknuts are used in boxes, these must be tight or a high resistance may develop in this ground line or no connection (an open, therefore useless ground) may result. See Fig. 2-5.

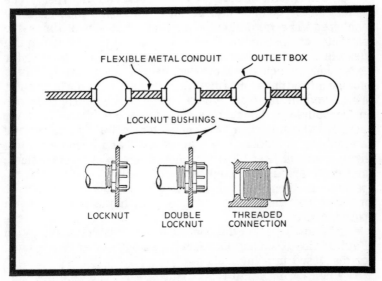

Fig. 2-5. Method of connecting conduit mechanically so that a continuous electrical connection is made through this metal. Wires are pulled from outlet box to outlet box.

Of course, all cables, conduit and boxes, etc. **shall** be fastened rigidly in place. The support spacings are 5 ft for ½ and ¾ in. conduit with 3 wires or less, and 6 to 7 ft if the conduit is 1 in. or larger.

Every run of raceway or conduit **shall** terminate in a box fitting. When additions are made to wiring and these require exposed (surface) runs, the cable shall follow the surface of the building or along boards such as baseboards. It shall be protected from physical damage by conduit, strips, or other means, and when passing through floors it shall be enclosed in rigid metal conduit or metal pipe extending **at least 6 in. above** the floor. In basements, runs may be made under joists, or through them, or on the side of a joist.

Nonmetallic cable can be used for extensions to existing outlets when it is run in dry locations, and is exposed (not run under plaster) in a home or office. This type cable is not to be used in unfinished attics, or basements where it may be exposed to dampness or corrosive vapors. In these areas, an airtight fixture and installation is required, using rigid tubing. If you run cable under plaster inside a wall, normally you will want to use and should use flexible BX, or an armored cable.

When using flexible conduit through which wires will later be pulled, be sure not to have any more bends than necessary. The code allows only 360 degrees of bending in one run between boxes, and not over a 90 degree bend anywhere. The radius of the bend should be large and the bends securely anchored. The minimum bend is at least 7 to 9 times the diameter of the conduit. There is always a temptation to use flexible 2-wire conductors such as used in floor lamps, for extensions, which somehow then become permanent. This is **not permitted** under the rules. Such wiring **must not** be used as a replacement for fixed type wiring with the appropriate size of 2-conductor cable such as we discussed earlier. The **lamp cord** or extension cord type conductors will fray and rot and are subject to damage by staples, etc., and become a most likely cause of fire. When it is necessary to have extensions, and these are used enough without change to be considered permanent, then replace the cords with flexible nonmetallic or metallic sheathed cable and save your home or office from possible fire.

It is now a part of the code that you must have a Ground Fault Indicator on any line serving an outdoor area. This will be discussed in a later chapter.

When wiring lamp type fixtures, etc., one common conductor throughout should have a specific identification to designate it as the ground wire. This color may be a white braid as against other color braids of the other conductors, or one may use a wire with a tracer in the braid when there are no other conductors with any tracers in them, i.e., all the other wires are all solid colors. Again the concept is to carry through the ground wire of the electrical system without interruption so it can be identified, and so that switches will not be placed in this line inadvertently and thus expose someone to electrocution because they came in contact with a hot wire and were then somehow grounded to earth through metal fittings, fixtures, grass, etc.

CONNECTIONS

There are certain approved ways of making connection between two or more electrical wires. In general, you remove the insulation from each wire, scrape until the metallic conductors are bright, twist them together tightly, and solder the joint if possible (using rosin-core solder only), then wrap thoroughly with plastic electrical tape until you have enough layers to equal the depth of the insulation removed. The manner of splicing is shown for various situations in Figs. 2-6 through 2-9.

Notice in Fig. 2-9 that if you must splice into an extension cord type of wire, that you remove the insulation at **two different points** in the wire. You make the connection or tap, and wrap tightly with electrical tape, taping first the joint or bare wire tap itself and then running the tape over both the bare wire tape and the insulated part of the wire.

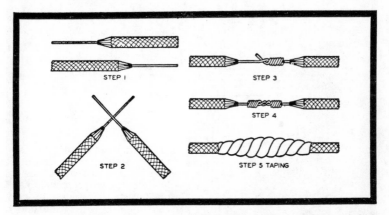

Fig. 2-6. Method of connecting two wires together. Remove the insulation with care so you do not nick the wire. Scrape the wire bright and clean. Use plenty of tape over the wire and on the insulation as in step 5.

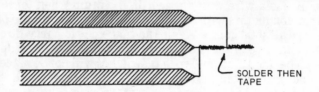

Fig. 2-7. Method of connecting three or more wires together.

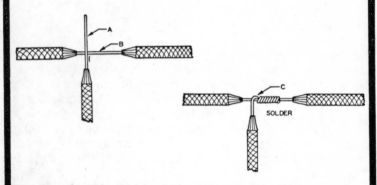

Fig. 2-8. Method of making a tap into a wire. The same size wire is used for the tap as is in the wire tapped. This connection is important in later work where we tap into a white wire to obtain a connection for an additional receptacle, or to extend some house or office wiring. The tap is used to make a connection to the white ground wire.

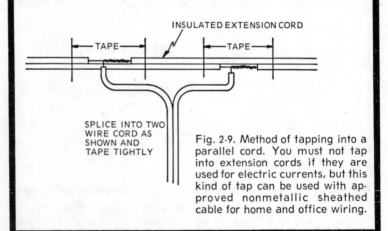

SPLICE INTO TWO WIRE CORD AS SHOWN AND TAPE TIGHTLY

Fig. 2-9. Method of tapping into a parallel cord. You must not tap into extension cords if they are used for electric currents, but this kind of tap can be used with approved nonmetallic sheathed cable for home and office wiring.

Basic Tools, Tests, and Simple Repairs

Chapter 3

I asked my wife what she knew about electricity. She gave me a superior glance and said, quite seriously, "Nothing!" "But you have a lot of appliances," said I. Her answer: "I plug them into the socket. If they don't work I call you!" And so ended our conversation on this subject.

This is typical of many people. Unless it is the television, and that in the middle of a favorite program, for some the next step is to call the electrician. Some people may check the fuses, and others may check to see if the neighbors' lights went out also. When it is a flash, a spark, and a sizzle at the plug end of an electric cord, then it may be a prayer of thanksgiving that all ended as well as it did while everyone stares at the charred ends of the extension cord, or the socket, now blackened by the flash flame.

In this book we are not recommending that "Electricity is for Everyone." Definitely, if it is not "your bag of gold," then follow your instincts and call the electrician. But if you do have a slight "bent" in this direction, then this and subsequent chapters will help to keep you on the correct procedural and practical paths to success in electrical projects.

TOOLS

Let's examine tools and things; see Fig. 3-1. There is some equipment which is necessary in testing, adjusting and repairing electrical circuits and equipment. These items constitute the fundamental tools which you need around the house. The basic equipment you need is 1 each of the following:

¼" blade screwdriver, long shank
⅜" blade screwdriver, heavy shank
⅛" **pocket-type** screwdriver
Long-nose pliers
Side-cutting pliers (called **dikes,** for **diagonal** cutters)

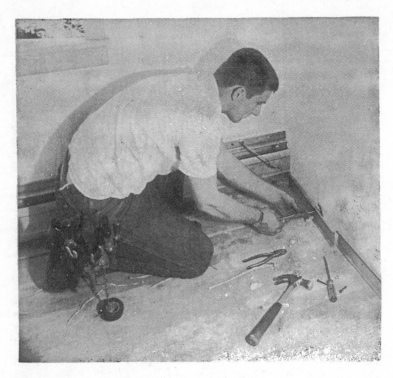

Fig. 3-1. The tools required are shown here. Drills and small metal and wood saws are also required. For conduit, a pipe wrench or open-end box wrenches may be used. (Wiremold Co.)

Electrician's knife
Roll of black plastic electrical tape
Hammer
Soldering gun or soldering iron, 100 to 250W
Roll of rosin-core solder (**never** use acid-core solder).

TEST DEVICES AND TESTING

The most common testing device is a **continuity tester** (Fig. 3-2) which may be made from a buzzer and battery, an old flashlight, or doorbell, or a lamp socket with long insulated leads. A 250V ac meter is nice, but not essential.

What's a continuity tester? It is an electrical testing unit which will cause electricity to flow through wires (a circuit) and show you by lighting a light, or ringing a bell, that the circuit is complete, and that the circuit will permit the electricity to flow through it. The unit operates on a minimum amount of electricity, as from batteries, so if there is anything wrong with the circuit, there will be no danger from the small current flowing from the tester. Once you know the circuit is okay by using the tester, **AND, OF COURSE, YOU HAVE TESTED THE CIRCUIT WITH POWER REMOVED,** then you can apply the higher voltage or regular power to the circuit and know that everything should be safe and sound and the things operated by the circuit will operate as they should. Figure 3-3 shows two important types of tests: the **continuity test** and the **short circuit** test.

In principle, a continuity tester is like a gallon jug of water. To see if water flows through a pipe, you pour water into one end and see if it comes out the other. With the continuity tester, after disconnecting incoming power, you connect one probe to one wire of the circuit, which might be thought of as putting electricity from your tester battery into this wire, and then you connect the other wire of your tester to the other wire of the circuit, to see if the electricity **comes out.** If it does, then the lamp lights, the bell rings, or the meter pointer deflects. If these things should happen and don't, then you must trace all the way around your electrical circuit to see if wires are broken, if they are shorted (insulation rubbed off and metal parts touching), or if there are defective switches, etc. along the line.

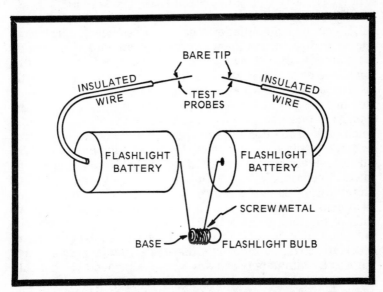

Fig. 3-2. A simple tester to determine continuity. If the circuit is complete, touching the bare tips across the wires will cause the flashlight to glow. There should be no 120V power on any circuit tested! To test the tester itself, touch the probe tips together—bulb will light.

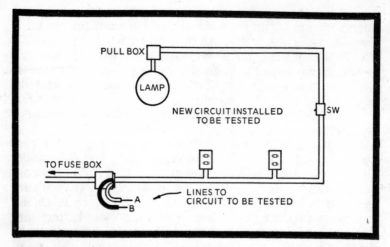

Fig. 3-3. When the wires are run in the conduit, AND BEFORE THEY ARE CONNECTED TO THE POWER FUSE BOX, the circuit may be tested. Turn switch (SW) on and test across A and B. If the circuit is okay, and the lamp bulb is good, the flashlight of your continuity tester will glow. You may also test for a short to ground by touching one probe to one line and the other probe tip to the conduit itself. If the flashlight now burns, you HAVE A SHORT. YOU MUST ELIMINATE THIS SHORT BEFORE YOU APPLY POWER! It helps to track it down if you open the switch (OFF) and test for ground on either side of the switch.

If you have an outlet or a pair of wires which carry electricity, and you do not know whether the amount of electricity is 120 or 220V, there is a way to test for this. Use two 120V lights connected as shown in Fig. 3-4. Be sure to remove just a little insulation from the "probe" tips of the test wires. It is a good idea to wear dry heavy duty rubber gloves during the testing also. Now, energize the circuit to be tested and very carefully touch both probe tips to bare spots on each of the two wires forming the circuit to be tested. You may have had to remove a little insulation from the **circuit** wires **before** you applied the power. The lamps will glow full bright on 220V and about half brightness on 120V. If it is a receptacle that you want to check for amount of voltage, then simply connect a plug to the two ends of the test wires at the probes, and plug this into the receptacle to be tested. You can also use this kind of tester to check any circuit to find out if there is a voltage present, or if there is a voltage between any wire and ground (or conduit, etc.). One note: You **cannot** check directly from a wire to **earth** ground with a tester such as this. Voltage can exist between, say, a cement floor and a wire which would shock you, but which cannot be shown to exist except with a **meter.** When you are in doubt, get an ac meter, 250V scale to test with.

SIMPLE REPAIRS

Most important of the repairs around the house (and office) which you can make easily, are those to lamps and devices which have plugs to plug into receptacles. The faulty items are old or frayed or broken electric cords or plugs. Examine all those you are using right now. Are cords bad, cracked from age, brittle, broken, frayed, etc.? The whole cord needs to be replaced if they are. Of course, if you have tried to make something light or operate and the cord seems to be okay, then the plug may be bad, the device may not be plugged into the wall socket tightly, the prongs may be bent and not make contact, or the receptacle may be faulty. Try another receptacle or plug to learn if this is so.

Replacing Cords, Plugs, Receptacles

One of the most common problems with a plug is that it isn't inserted properly in the socket. Sad but true. So if a device doesn't work, push the plug in tighter and see if that helps. If it keeps coming loose, install a new receptacle.

Next, the **prongs** on the plug may have become **slightly** bent toward each other, or away from each other, Fig. 3-5. In either case it may be that even when plugged in firmly they **will not** make contact with the mating metal part inside the socket. In this case, you may have to pry the prongs apart slightly or push them together slightly and try the device. You will find that with plugs and sockets which have had much use, this may work. Again, replacing a worn receptacle may be required, Fig. 3-6.

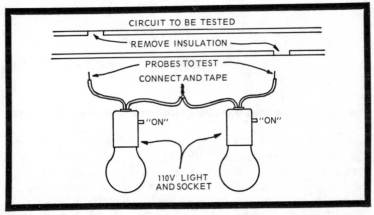

Fig. 3-4. A tester for higher voltage circuits. You can test up to 220V with this unit.

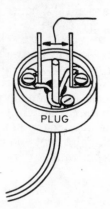

SPACING TOO CLOSE (BAD)
SPACING TOO FAR (BAD)
ADJUST

LOOSE PRONGS (BAD)
REPLACE PLUG

CORRODED PRONGS (BAD)
REPLACE OR POLISH.

PLUG

Fig. 3-5. The proper way to wire a plug is to bring the wire around the plug blade. The screw should tighten the wire, not loosen it.

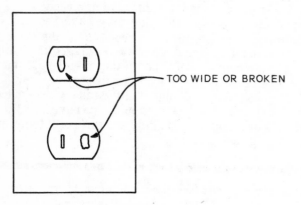

TOO WIDE OR BROKEN

Fig. 3-6. Examples of bad receptacles. Enlarged holes mean bad connections. The receptacle should be replaced.

In some cases, though, the **plug** is bad, or the cord right at the point where it goes into the plug may be frayed and broken and dangerous. For this situation, a new plug is desirable. Cut off the old cord about 3 in. back from the bad end, Fig. 3-7. Carefully insert a knife between the two parallel wires and cut the insulation slightly. You can then just pull the two ends apart for a distance of about 3 in., run the wires through the

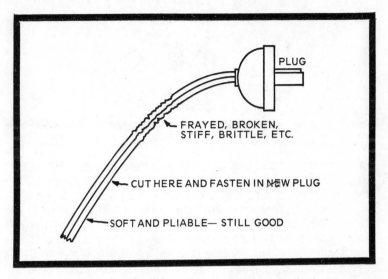

Fig. 3-7. Replace the plug when the cord is bad. See text.

plug hole, and tie. Now, carefully cut the insulation around each wire end 1 in. back from the end and strip it away (Fig. 3-8). Be certain that you can see the metal wires. These will probably be stranded and will be bright coppery looking if you have all the insulation stripped away. Now twist each end tightly. You have pulled the two wires through the new plug. You may desire to tie the underwriter's knot to keep the cord from putting strain on the electrical screw connections. The knot is shown in Fig. 3-9, and is tied **after** pulling the cord through the plug hole. Now, pull each end AROUND the plug pin, and tighten it under the screw head. Be sure the direction

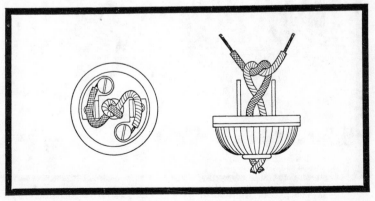

Fig. 3-8. Use of a single knot to hold electric cord in the plug.

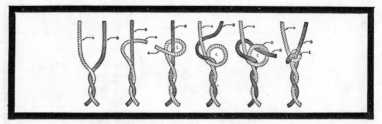

Fig. 3-9. The underwriter's knot. Try it instead of the knot of Fig. 3-8.

of the wires around the screw is as shown in Fig. 3-10, so that as you tighten the screw it tends to **pull** the wire tighter, not loosen the wire. Cut away all remaining strands.

Of course, there is always the possibility that the socket or receptacle itself may have finally worn to a point where it just won't make contact any longer. When we test this receptacle (or socket), we can wiggle the test probes of our two lights around to such a position that we **can** make electrical contact and thus we believe that the socket is okay. This may not be true. If the appliance or light doesn't light, but its plug and cord are okay, **and** you can get an indication with probe testers that the receptacle has power applied to it, then it may be necessary to **replace the receptacle** (Fig. 3-11). To replace a receptacle, here's all you do:

1. Shut off the current at the main circuit protective panel.

2. Remove cover plate and remove receptacle from box.

3. Disconnect black and white wires from terminals. (Also green, if the device is grounded.)

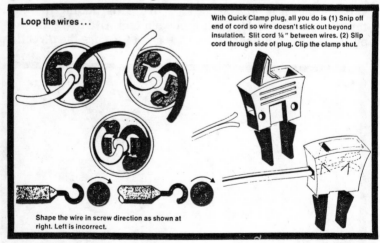

Loop the wires . . .

With Quick Clamp plug, all you do is (1) Snip off end of cord so wire doesn't stick out beyond insulation. Slit cord ¼" between wires. (2) Slip cord through side of plug. Clip the clamp shut.

Shape the wire in screw direction as shown at right. Left is incorrect.

Fig. 3-10. The proper method to wire a plug is shown. Also the way to connect the Quick Clamp plug. (G.E.)

In most receptacle installations you'll find two sets of black and white wire pairs—a "coming in " set, and a "going out" set. This means that the device is "through wired" so that the circuit continues beyond that particular outlet.

4. Connect the black wires to the brass terminals, and the white wires to the silver terminals. (Most receptacles are designed for through wiring and have double binding screws or terminals to facilitate connection of the in-and-out sets of wires.) **Never connect black to white wire, or to same terminal screw.** If the device is grounded, connect the grounding wire to the green terminal.

5. Use continuity tester; check for continuity, shorts, grounds.

6. Mount receptacle in box and replace cover plate.

7. Restore current at circuit protective panel.

The first step was to insure that the electricity is **off.** To determine this, connect a lamp on the same circuit, then turn off **all** master switches or circuit breakers at the fuse box.

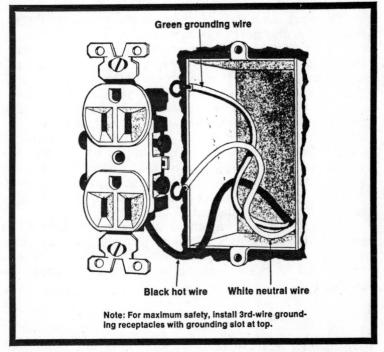

Green grounding wire

Black hot wire   **White neutral wire**

Note: For maximum safety, install 3rd-wire grounding receptacles with grounding slot at top.

Fig. 3-11. Replacement of a receptacle is easy if you disconnect the power first! Note the green grounding wire. You socket receptacle wiring may not have this line. You can add it by simply connecting a wire from the grounding terminal on the socket to the metal box itself.

NOTICE THAT WE SAID THAT **ALL** SWITCHES ARE TURNED **OFF**. This may be a little inconvenient, but if you aren't sure which one actually operates that particular receptacle, then MAKE SURE by the above procedure. (Even if you must work by flashlight, and it is inconvenient, at least you'll still be around to read my next book!) **Now** you can unscrew the receptacle plate, unscrew the two screws (top and bottom) which hold the receptacle in place in the box, pull it forward, unscrew the side screws to the two wires and replace with a new receptacle of the same type. There are receptacles which permit you to merely insert the ends of a wire into two small holes to make connection. Personally, I don't care for this type. The kind with screws on the side so that you can tighten down on the wire and be POSITIVE of a connection to the socket is much more satisfying. Anyway, now that you've done the job and replaced everything, you can turn all fuse panel switches on again.

Switch Troubleshooting

The next main item to check is the switch on the appliance. You may not be able to repair it, but you can check to see if it is at fault, or if an element in the applicance is at fault. The way to do this is to first check your outlet to be sure you have electrical power. Plug in a lamp which you know works. If it turns off and on as it should, then you can say you have electrical power at that socket. Next, check the appliance plug and cord. The cord should be checked carefully to see if it has a broken wire in it. Of course, do **not** have it connected to electrical outlet when you make this check. Flex it. Sometimes you can feel the broken place in the cord if it exists. If the cord and plug are okay and you do have electricity at the socket, then the appliance itself is at fault and you will probably have to send it out to be repaired.

It's easy to replace a light socket which may have a bad switch in it. Refer to Figs. 3-12 and 3-13 to see how this is done.

Once in a while a wall light switch will go bad and have to be replaced. Make sure power is off, and follow the procedure in Fig. 3-14.

Cords and Safety

Consider again extension cords. They are simple and necessary but if abused can cause much damage. Remember, the extension cord is not designed to stand rough usage. It is supposed to be just a **temporary** means of extending an

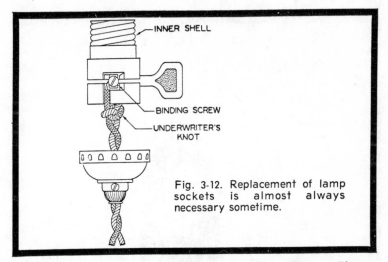

INNER SHELL

BINDING SCREW

UNDERWRITER'S KNOT

Fig. 3-12. Replacement of lamp sockets is almost always necessary sometime.

electrical connection to a light or small appliance. If you kink it, crush it, tie knots in it (to shorten its length) you may cause a short circuit and fire or get shocked when you handle the "birds nest" in the cord. Where can you get killed? The best places are in a damp basement, the laundry where there is a good damp connection to the floor and in the kitchen or bath where you can hold your cord and also might touch the water faucets at the same time. BE CAREFUL!

When using extension cords, coil them in LARGE loops. Do not bend them unnecessarily. Don't run furniture wheels over them, don't lay them in home traffic passages. People may trip and fall, or damage the cord. Don't use them for **permanent** extensions to existing wiring! (More on this in Chapter 5.)

Another thing: Some electrical cords are specially constructed for specific appliance uses. If you replace one at random, without consideration for the kind of equipment it will operate, you **could** be asking for trouble. An extension cord suitable for a light may not be suitable for an air conditioner or heater. Try feeling the cord with your hand; grasp it firmly when it is conducting electricity to an appliance or other device. If it is **warm** or **hot** to your hand, then it **does not** have sufficient capacity for the current which must flow. Its wires are too small! You need a cord with a larger current carrying capacity. This heat can cause the insulation to break, dry out, crack, even **melt**, and you'll have resultant trouble.

When considering **appliances**, do not replace a worn and frayed cord (which came on the appliance) with just any other which you may have on hand. Always be sure that you replace

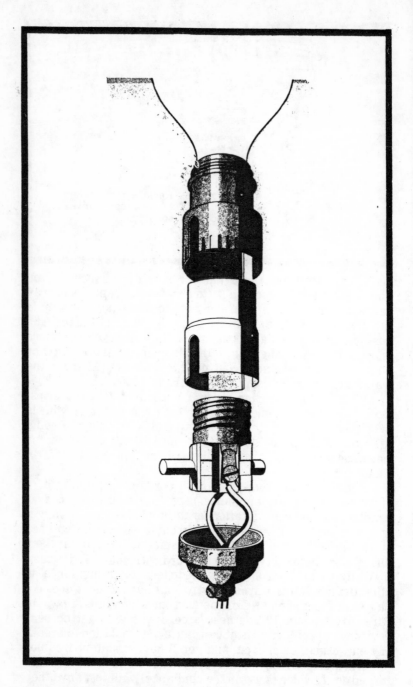

Fig. 3-13. Another type of lamp socket with a pushbutton switch. (G.E.)

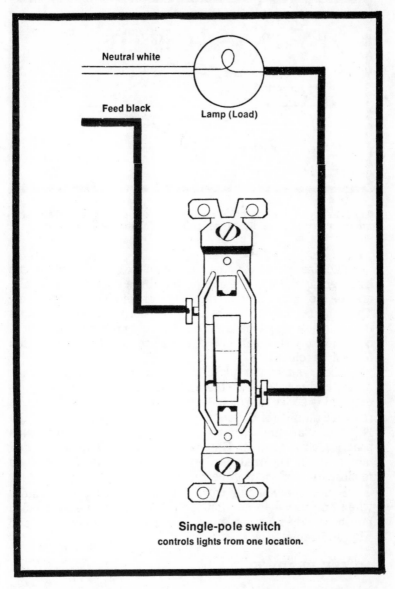

Neutral white

Feed black

Lamp (Load)

Single-pole switch
controls lights from one location.

Fig. 3-14. Replacement of the house or office light switch is also often required. The insulation on the line from the switch to the light is normally red. (G.E.)

with exactly the same kind that was originally on the unit. You might even take the old cord with you when you're shopping for a replacement.

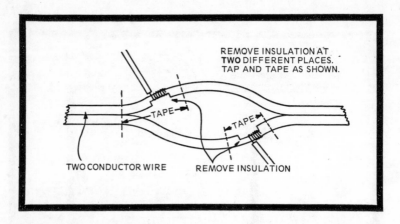

Fig. 3-15. When tapping into a 2-wire cable, use spaced taps. Tape each tap independently and then tape the whole section carefully with plastic electrical tape.

One final precaution on the use of extension cords in the home garage. Please be aware that there may be chemicals, grease, etc., from the car, mower, etc. which get on the floor. These in turn may get on the extension cord and eat away at its insulation. Always be alert to the possibility of chemical corrosion of your electrical extension cords. Check them often to avoid accidents!

When you contemplate the permanent wiring in your home, and you think of the addition of outlets, you may be tempted to splice into an extension cord in order to obtain the required number of outlets. If you are so tempted, DON'T DO IT! The National Electrical Code expressly forbids this practice as a hazard. Extension cords for appliances and lamps are to be used as they are purchased only, and no modification of them is permitted.

Speaking of splicing cords together: There are occasions when one needs to do this for other than power extension purposes. When a tap is made it must be done correctly and Fig. 3-15 shows how.

METER READING

Let us digress for just a moment. What does your electricity cost you? Can you read the meter? Can you estimate the cost of operating an appliance which you might want to buy? It is easy. First let's read the meter. See Fig. 3-16.

Now let us calculate the cost of electricity; just follow my calculations:

$$\frac{W}{1} \times \frac{\cancel{c}}{kW\text{-}hr} \times \frac{0.001\ kW}{1W} = \cancel{c}/hr \text{ (operating cost)}$$

This says that if we know the wattage of a device, and we know how much the electricity costs us per kilowatt-hour (we can call the electric company), then we can calculate the operating cost in cents per hour. For example: Let us examine an electric heater which uses 1000W continuously when it is on. Let us assume that the cost of electricity where you live is 5 cents per kilowatt hour, then:

$$\frac{(1000)}{1} \times \frac{(5)}{1} \times \frac{(0.001)}{1} = \begin{array}{l} 5 \text{ cents per hour} \\ \text{operating cost} \end{array}$$

Sometimes it is valuable to know these things because the electric company may send you an estimated bill time after time and you may not actually be using the amount of electricity they have estimated. For example, it was my personal experience to have received an electric bill of $35.00 each month for 2 months in succession when the usual bill had been about $25.00. It was noted (by the code on the bill card) that the bill had been **estimated**. A call was made to check this and it was explained that this estimation had been based on the bills of **other houses** in the neighborhood close by. The reason given for estimating was that since the back yard gate was locked, the meter reader couldn't get to the meter to get a reading. Since that time, with the cooperation of the electric company, I read the meter and sent them a form card (which they supply) with the readings on it, and we are all happy!

Speaking of costs, here is a way to save some money. The manufacturers of light bulbs make some which are rated at

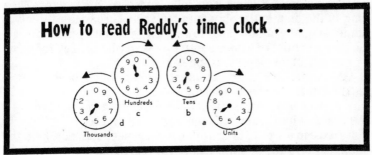

How to read Reddy's time clock . . .

Fig. 3-16. To read the meter, start with dial (a). When a hand points between numbers, read the **lower** number. Here the dials read: (a) 6, (b) 4, (c) 9. Dial (d) would read 4 IF dial (c) had passed zero. Since dial (c) had NOT passed zero, then we read dial (d) as 3 and not the value of 4 which it seems to be pointing to. The reading then is 3946. Notice that you place the numbers down from right to left as you read them.

125 to 130V instead of the usual 110 to 120V. Ask for them at your store. This type will last much longer then the regular bulbs and they do not cost but a small amount more than the usual type. The light output is not noticeably diminished either. It's a fine way to save, especially if there are lots of lights in the house, and children who must use them long hours to study, or who forget to turn them off.

APPLIANCE REPAIR PRECAUTIONS

Repairs to appliances must come under a separate book heading. This is a subject too large to include here, but there are a few things which can be relatively easily checked and fixed which are appropriate to this text. One of these is the replacement of an **electric range element.**

When the element fails to function it is due probably to one of three causes—the switch is not making contact, the connections under the range surface to the element are faulty, or the element is burned out. Since there is usually a light associated with the switch for that element, the fact that this glows is a reasonably good sign that the electricity is going through the switch. We have examined several cases, however, where the connecting socket to the element itself has become heated, even charred, and has burned a connecting wire so that it no longer makes contact with the socket. To examine this situation, TURN OFF POWER AT THE FUSE BOX, then raise the element, unscrew the pivot fixture, and pull the element and its wires onto the range top. You will see the loose wire and you can determine where it is normally connected. You may have to obtain a new receptacle socket, and then clean all wires carefully and reconnect them. Then plug in the element prongs into the socket, refasten the pivot hinge in place again and lower the element back into its normal position. Use your continuity tester and check for short or ground before applying power.

A connection at the element socket which becomes the least bit corroded becomes a bad electrical connection. This in turn causes heat, and sufficient heat can be generated to burn the wires in half, char the socket insulation, etc. It isn't a bad idea, once in a while, to examine each socket connection to the elements on your electric range to insure that they are clean and tidy, that no spilled food or liquid has started a corrosion process.

Look at your electric oven. There is, in some, a dangerous situation which can exist, especially if the oven is a type which you must clean, and which has a small light mounted toward the back which is not enclosed in an outer glass protective cover. These lights get very hot (of course they are designed to withstand the heat of the oven), but when you clean, you MAY rub across the bulb with a damp and normally lukewarm or cold cloth. The bulb may burst! There will be "live" electrical connections exposed (the wires inside the bulb). You will probably jerk back and may hurt yourself just in your physical reaction to the explosion, and you MAY touch the wires in doing so and get a bad shock also. BE CAREFUL when cleaning ovens that have light bulbs in them which are not protected by the extra glass cover. NEVER touch them, when lighted, with a damp cleaning cloth! Your best protection, of course, is to turn off all power to the electric range and let it cool down completely before you work inside the oven. But if you have a shattered bulb and have to replace it, then put on rubber gloves, BE SURE ALL ELECTRICITY IS OFF, try to twist the remnants of the bulb out of the socket with the glass stem (which usually remains intact). Sometimes you may have to get a small pair of pliers and grasp the stem gently and twist. Once it's removed, then you can put in the replacement and when you do this, DO NOT SCREW IT IN AS TIGHT AS POSSIBLE. Be prudent. Screw it in so it makes a good contact but no tighter than necessary. If it is screwed into the socket very tightly, it will pose a problem on removal if it does ever shatter. If the bulb burns out and you have to remove it and it is stuck, wrap it with masking tape and then try unscrewing it. You'll have a firmer grip and may be able to do the job easily.

ELECTRICAL INTERFERENCE

Let's spend a moment talking about electrical interfence which may be generated within the house wiring itself. In many homes there are dimmer switches used in place of the off-on switches. These permit you to vary the light intensity gradually and are ideal for nurseries, or for living rooms where you want to have low level lighting to increase informality or heighten romantic atmosphere. But these same dimmer switches, depending upon how they operate, may be a source of electrical interference for your radio or even the TV. There is one sure way to check this. Turn the radio on and increase the volume to a reasonably high level. Now turn on the dimmers one at a time, meantime listening and tuning the radio dial throughout the full range each time you turn another light on. You will identify the source of the interference, if it exists, in this manner and it may surprise you to find that it belongs to a switch located all the way across the house—one that you would never normally suspect! You may replace it with another, different make, or sometimes even the same make, and clear the trouble.

Of course, you are familiar with the interference patterns, dot-like streaks across the TV screen, in bands, when the dishwasher or the garbage disposal, etc., are running. If you have a problem with this and you cannot rearrange the operating schedule of the dishwasher or garbage disposal, etc., then visit your local radio store and buy a line filter. This is a small unit which will plug into the wall socket of the appliance, and the appliance cord is plugged into it, and the interference will be trapped out.

Basic Illumination Engineering

This chapter is a wee bit technical, but it is about a pleasant subject, proper lighting (Fig. 4-1). We need to understand some of the whys and wherefores of lighting so that we will be able to accomplish what we want to accomplish with light, and know how and why we are doing what we do. There are certain rules to be followed to have a beautifully lighted home or office or grounds, and yet, while beautiful, still to be fully utilitarian. We will examine the concepts of these rules. We will need to know definitions so we can speak intelligently and in the language of the interior decorator or lighting engineer. Let us begin, then, with an examination of the sensitivity of the eye to various colors of light, Fig. 4-2.

Notice that the eye is most sensitive to **green**. This means that if you had all colors displayed before you at the same intensity, you would think that green was **brighter** than the others. Violet would seem the dimmest, and dark red would appear less intense also.

The question immediately arises, "Why, if green is most intense to the human eye, does the highway department use yellow or orange on road signs?" The answer is **contrast**. If you use green against a green or brown background, the contrast is less than the yellow or orange. Notice that the yellow is next in intensity brightness to green. So this choice of colors is best for intensity **and** contrast. A little later on we will examine some color effects in lighting. It will be well to remember this chart at that time.

DEFINITIONS

There are certain terms associated with lighting which you should know, to be able to discuss this science. These terms are:

    a. **Illumination**—The light flux on a surface.

    b. **Lumen**—Amount of light flux which produces 1 foot-candle of illumination on EACH POINT of a 1 sq ft area. Or a

Fig. 4-1. Proper lighting of a breakfast nook can create a very pleasant atmosphere. (G.E.)

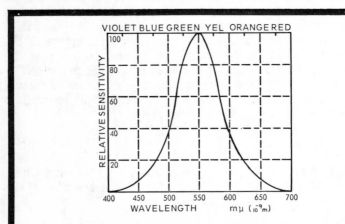

Fig. 4-2. The relative eye sensitivity for an assumed standard observer for normal levels of illumination.

Table 4-1. Watts vs Lumens for General Service Lamps	
WATTS	**LUMENS**
10	80
25	270
60	840
100	1640
150	2700
250	4400

lumen is a unit of light intensity over the entire 1 sq ft surface from the one candle. Think of the page of a book 12 by 12 in. held in a slight curve so that the light is the same from a candle on all points on the page. You have one lumen of light on the page. This level of light would make reading difficult, wouldn't it? (There is another term, included for academic interest only; the **candela**. The candela is the **luminous intensity** over the entire 1 square foot area. There is a little difference between the illumination and luminous intensity. The first relates to the amount of light or **brightness**, the second relates to the amount of light flux which is flowing to the surface. The eye requires a luminance of 0.3 candela per square meter to see colors.)

Please notice the light bulb rack the next time you are in the grocery store. You will see, on the boxes of bulbs, a notation as to the amount of lumens the bulb will produce. For example, a 100W bulb will produce 1640 lumens. See Table 4-1.

c. **Foot-candle**—(first definition)—is lumens per square foot. If you divide the lumen value by the square feet of room surface area, you will determine the average foot-candles of illumination. See Fig. 4-3 for the distribution of light in candlepower from an ordinary light bulb.

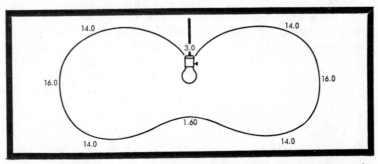

Fig. 4-3. Candlepower illumination distribution from an ordinary incandescent lamp.

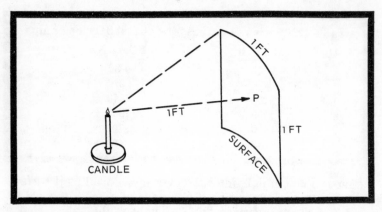

Fig. 4-4. The concept of a candlepower of light. The light intensity at point P is 1 candlepower.

    d. Foot-candle— (second definition)—This is actually the amount of light falling on a point 1 ft away from a candle in a horizontal direction, Fig. 4-4. Sunlight on a clear day at sea level is about 10,000 foot-candles per square centimeter. On a cloudy day this decreases to about 1000 foot-candles.

    e. Photometer—This is an instrument like a camera lightmeter which measures luminous intensity in candlepower. Visit your camera store and ask about such a meter and examine it.

    f. Foot-candles—(third definition)—This is actually the **candlepower of a lamp considering a distance** (in feet). For example, if the lamp had a candlepower of 10 candles, and you wanted to know the foot-candles at 5 ft, you divide

$$\frac{10C}{25 \text{ ft}^2}$$

to get the foot-candles at that distance. Remember that light intensity varies inversely as the square of the distance.

    g. Lambert—A unit of brightness. One lambert is obtained when a surface radiates or reflects one lumen per sq centimeter; 2.5 (approximately) centimeters equal one linear inch. The brightness of one candle per sq in. is 2.05 lamberts. Many times, illumination engineers will use lamberts as a figure of brightness of the illumination to be achieved. See Table 4-2 for some lambert values. One lambert=295.7 candles per sq ft. One foot-candle=0.00338 lambert.

    There are many kinds of tables which show the foot-candles of light required, or recommended for various tasks and situations. Let us now examine one such table. Table 4-3 shows the relative ratings of incandescent (regular light

Table 4-2. Lambert Values

CONDITION	Foot-Lamberts
Ground on sunny day	100
Snow-full sunlight	5000
Overcast day	10 to 30
Clear moon	0.01
Twilight	1.0

bulbs) in lumens as compared to fluorescent tubes of low wattages. In Table 4-4 we see some examples of lighting levels as recommended by the Illumination Engineering Society (IES) Handbook.

CALCULATING LIGHTING LEVELS

All of these levels of lighting are to emphasize the planes and curves of structures, or to provide light to see and work or play, and to establish moods. Let us now perform some simple calculations with foot-candles and lumens. Your local

Table 4-3. Comparison of Incandescent and Fluorescent

INCANDESCENT LAMPS

WATTS	LUMENS
10	80
25	270
60	840
100	1640
150	2700
250	4400

WHITE FLUORESCENT

WATTS	LUMENS
4	100
8	330
15	760
20 (circular type)	1550 (may be as low as 690)
30	1900
40	2500
90	5150

Table 4-4. Foot-Candle Levels Recommended on Task (IES Handbook)

AREA	FOOT-CANDLES
Entrances	10
Living rooms	50 to 70
Bath	30 to 50
Kitchen	50 to 70
Makeup vanity	50
Offices	100 to 200 ·
Gameroom	30
Sink	70
Range (work surface)	50
Laundry—iron room	50
Study—read-write	30
Desks, study	70
Sewing dark material	200
Other fabrics, sewing	100
Shaving—makeup	50
Stairs—entrances	10 (may be higher if people older)
Living rooms	10 to 50 variable desired
Family rooms, dens	10 to 60 variable desired
Kitchen—laundry, etc.	30

OUTSIDE ILLUMINATION

Garden	0.5
Paths	1.0
Background fence, trees, etc.	2.0
Flower beds, rock gardens	5.0
Trees, shrubbery	5.0
Focal areas, large	10.0
Focal areas, small	20.0

OFFICE AREA ILLUMINATION

Design areas and drafting	200
Accounting desks	150
Regular office task areas	100
Steno desks	70
Well printed material reading areas	30

photographic shop can show you some lightmeters which probably have a foot-candle scale, or table by which you can convert a reading into foot-candles. You can then measure the light levels in your home, office, or work shop.

Since most lighting engineers, decorators, builders, etc. talk in terms of foot-candles and lumens, we relate these as follows:

$$\text{foot-candles} = \frac{\text{lumens (times a degradations factor due to reflections, dirt, etc.)}}{\text{area in square feet}}$$

Example: Suppose you have a table top 2 by 2 ft. Suppose that you have it lighted by a lamp which produces 50 lumens and that this light is degraded 10 percent by dirt on the bulb and by reflections which tend to cancel the light rays. Then the foot-candles per sq ft on the surface is:

$$\frac{(50) \quad (0.1)}{(4 \text{ sq ft})} = 1.25 \text{ foot-candles}$$

Suppose that 40 foot-candles are required for proper illumination on a desk. The equation to solve for lumens still assuming the 10 percent degradation factor is:

$$\text{lumens} = \frac{(\text{area, sq ft}) \quad (\text{foot-candles})}{(\text{degradation factor})}$$

$$= \frac{(4) \quad (40)}{0.1} = 1600 \text{ lumens}$$

From Table 4-3 we see that a 100W bulb will do the job.

So now we have some idea of the relationship between foot-candles and lumens and we will hear these terms or read about them whenever we consider lighting anywhere. Look for the lumen numbers on the light bulb box at your store. Manufacturers make charts showing the foot-candle distribution from their fixtures as in Fig. 4-5; ask for them when you buy fixtures.

Next we will define more terms which are important to people in the lighting business.

Room Cavity—This is the volume formed by an upper plane through the light fixtures, down to the work level and bounded by the walls, Fig. 4-6.

Ceiling cavity—This is the volume which is formed by the plane through the light fixtures and the ceiling proper, bounded by the walls, Fig. 4-7.

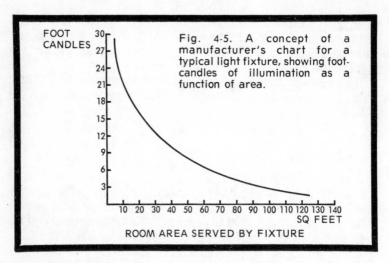

Fig. 4-5. A concept of a manufacturer's chart for a typical light fixture, showing foot-candles of illumination as a function of area.

Floor cavity—This is the volume formed by the plane through the work surface and the floor and bounded by the walls, Fig. 4-8.

Of course, if the fixtures are in the ceiling then there is no ceiling cavity as such. The lighting in each of the cavity areas is, or may be, different. The reflectance of the surfaces in each cavity may be different, and involved calculations are made (using computers) to insure the correct light in each cavity.

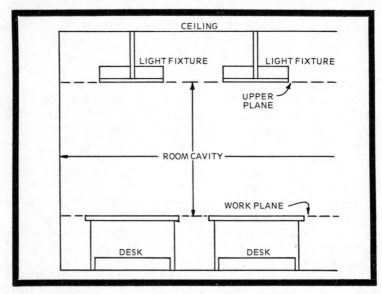

Fig. 4-6. Illustration of **room cavity**.

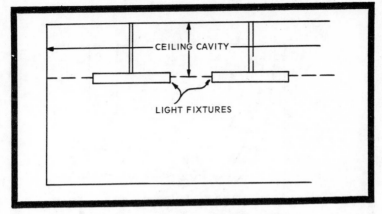

Fig. 4-7. Illustration of **ceiling cavity.**

When we consider the amount of illumination, we need to be aware of the factors which **reduce** illumination intensity AFTER the lights and fixtures have been installed. These are:

a. The age of the **fixture**, which may degrade the paint or finish which gives a reflectance.

b. **Dirt**, of course, will reduce light. The dirt accumulation of the fixture itself and on the walls, etc.

c. Burned out bulbs in multiple bulb fixtures may go **almost** unnoticed, but do reduce the light intensity.

d. **A reduction of the supply voltage** to the house—or perhaps more accurately, in homes which have their wiring for a specific load and then the owners **add** more and more electric devices to the circuits until circuits are loaded very

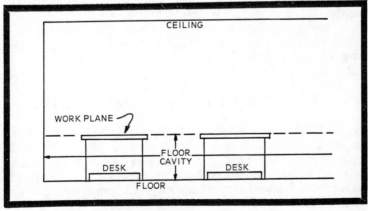

Fig. 4-8. Illustration of **floor cavity.**

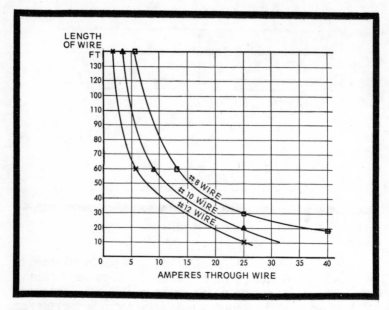

Fig. 4-9. This graph relates the length of a wire line to the amount of current which can be drawn with not over a 1V loss. Notice that various sizes of wires have different current capacities over the same distances.

heavily. This **may** cause a reduction in voltage. In neighborhoods where pole transformers were installed with a given rating based on a given load, years go by and the loads in **all houses** increase and the voltage may drop. You can ask your electric company to check your line voltage if you have trouble or dimness or appliances don't seem to function exactly as they should. The electric company will check your line voltage during a 24 hour period without charge, usually.

e. **Lamp age.** It may still be burning, and has been for a couple of years. Lamps and bulbs do lose their intensity with time. See if a new bulb of the same wattage isn't brighter in that socket!

APPLIANCE LOADING VS LOW VOLTAGE EFFECTS

We have mentioned wiring, indirectly, when we discussed the loading effect of many appliances in the home. What happens is that if a small size wire was used to wire a house, then as the load increases this wire develops a **resistance** to the increased flow of current and causes a **voltage drop** or loss of voltage across its length. This loss is what makes your lights dim or the appliances work marginally; e.g., TV is unstable, electric heaters don't heat as they should, etc. There are tables and graphs which relate the amount of voltage drop or loss of voltage to the length and size of wire. One such graph is shown in Fig. 4-9.

The way in which you read this graph is that if, for example, you have a line consisting of two No. 12 wires which have an unbroken distance of 50 ft, you locate the 50 ft line on the left vertical scale, move to the right to the curve for the 12 size wire and read down—you will see that you can draw no more than 7.5A if you do not want to have more than a 1V drop (or loss) in this line. Remember the current you would draw can be calculated by dividing the total **wattage** on the line by the voltage (120). In this case the total wattage is limited to: 120V x 7.5A = 900W.

WIRE SIZE VS ILLUMINATION LEVELS

There is another kind of chart which relates to the wiring in a home. This one relates to the wire size (after some calculations) required for the various types of fixtures or lights considering the amount of foot-candles illumination desired. See Fig. 4-10. This graph shows, for example, that if you want an average illumination of 45 ft-candles, then you must plan for a wiring system which will accommodate at least 1.75W per sq ft if you use general (or diffuse) fluorescent lights, 2.75W per sq ft if you plan to use semi-indirect fluorescent lighting, 4W per sq ft if you plan to use indirect fluorescent lighting, and 5W per sq ft if you plan to use direct-indirect type incandescent lighting.

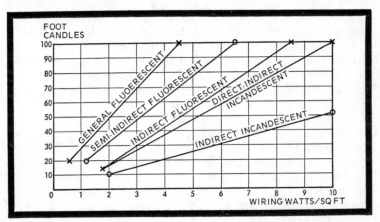

Fig. 4-10. Planning chart for development of home or office wiring. The circuit capacity is related to type of illumination used. See text.

To look at the current, and from this to determine the wire size to be used and number of 20A circuits necessary, let us imagine a room which is 25 by 15 ft. The total floor area is then 375 sq ft. If we planned to use an indirect fluorescent lighting system, we would use 4W per sq ft so the total wattage would be 1500W. One 20A circuit at 120V will provide 2400W and so only one circuit would be required. We could use some receptacles also on this line. The wire size would be No. 12 or No. 10 depending on length.

CALCULATING TOTAL ELECTRIC LOADING

There are some engineers who recommend the following: for each **500 sq ft** of house area (calculated on **outside** dimensions) plan **one** 120V, 20A circuit. If there is any fractional footage left over, then plan an additional circuit for this. The total number of circuits times 2000 will be the total home wattage capability load, at 4W per sq ft. Add any special loads to this. Two No. 12 wires are okay for each circuit, if runs aren't over 20 ft (see Fig. 4-9), considering the 2000W per-circuit value. At 2000W per circuit, the amperage is 16.7A. Fuse each circuit at 20A.

PRACTICALITIES

While at first one may think that fluorescent lights run cold, a moment's reflection will indicate that this may not be true. It is a fact that fluorescent lights get hot in operation. Some larger sizes, 40W, for example, can get too hot to hold comfortably in your hand. Another item of importance, the ballast transformer which is contained inside the fixture gets very hot also and so radiates heat to the fixture itself. This means that when mounting the fluorescent lights, adequate precaution must be taken to insure an air space around the fixture, when it is mounted on a plaster or other type of fireproof structure or wall or ceiling.

Installation of a fluorescent fixture to replace an incandescent one is not a particular problem. Like the regular light fixture, there are two wires, black and white. The white goes to the white and the black to the **red** in the box where the light was installed. Mounting of the fixture to the box can be accomplished by drilling holes in the fluorescent base to match the electrical conduit box ear lugs; then screw this part of the fixture in place with at least a **half inch** air space between this base of the fixture and the wall. Of course, in very small units (of 20W or less) the heat generated may be low enough so that if you have a plasterboard wall you can mount the fixture directly to it.

Next we examine some general lighting requirements for various home rooms of types and sizes in Fig. 4-11. It is appropriate here to mention the types of dimmers which can be used in the home or office to control the light levels. The General Electric "Specifier" dimmer can control either the incandescent or the fluorescent type of light. The control is shown in Fig. 4-12. A typical installation is shown in Fig. 4-13

	FIXTURES	STRUCTURAL
AREAS	**Ceiling-Mounted or Suspended**	**Valance/Cornice Wall Bracket***
LIVING	(light widely distributed)	
Living Room, Family, Bedroom Small (under 150 sq. ft.)	3- to 5-socket fixture, total 150-200 watts	8 - 12 ft.
Average (185-250 sq. ft.)	4- to 6-socket fixture, total 200 to 300 watts	16 - 20 ft.
Large (over 250 sq. ft.)	1 watt per sq. ft. and 1 fixture per 125 sq. ft.	1 foot/15 sq. ft.
UTILITY		
Kitchen, Laundry, Workshop Small (under 75 sq. ft.)	150 watt incandescent or 60 watt fluorescent	Use single row of fluorescents on top of open-to-ceiling cabinets—or in soffit extended 8"-12" beyond cabinets
Average (75-120 sq. ft.)	Incandescent: 150-200 watt or Fluorescent 60-80 watt	or
Large (over 120 sq. ft.)	Incandescent: 2 watts per sq. ft. or Fluorescent: ¾ to 1 watt per sq. ft.	5 to 6 watts/sq. ft. in fluorescent for luminous ceiling

RECESSED

Directional Flood Type**	Non-Directional Square or Rectangular Boxes
	(Recreation or Family Rooms)
	Four 75 watt incandescent
Four R-20 50 watt units	Two 40 watt fluorescent
Five-Eight 75 watt R-30 or 150 watt R-40 units	††Four 100 watt incandescent Three 40 watt or Four 30 watt fluorescent
One 75 watt R-30 unit for each 30 to 40 sq. ft.	Incandescent: ††One 100 watt or 150 watt per 40 - 50 sq. ft.
	Fluorescent: Two 40 watt or three 30 watt or six 20 watt per 100 sq. ft.
Not suitable for general lighting in these areas	††Two 150 watt incandescent or two 40 watt fluorescent
	††Four 100 watt incandescent or two 40 watt fluorescent
	Incandescent: ††One 100 watt per 30 sq. ft. or ††one 150 watt per 40 sq. ft.
	Fluorescent: Two 40 watt or three 30 watt or six 20 watt per 60 sq. ft.

*Or use equal length of wall lighting with recessed wall washers or floodlights.   ††Minimum Size: 10", 12" or 14" preferable.
**Should not be located above heads of seated persons; may be 18" or more in any direction from this point.

Fig. 4-11. Minimum general lighting needed for various room types and sizes. Various types of lighting systems are also considered. (G.E.)

Fig. 4-12. Typical G.E. control for the "Specifier" dimmer which can be used on incandescent or fluorescent type lights. The control pictured is for incandescent only, 2000W maximum.

but notice that the installation **circuit** in this figure is for the incandescent type of light. The circuit installation for the fluorescent type light is shown in Fig. 4-14.

As you know, it is possible to get some interference from any kind of dimmer. The following recommendations from G.E. may be of value in eliminating this kind of trouble, especially if the dimmer units are used in offices or schools or in rooms where **public address systems are used.** At home, you will find sensitive radios and hi-fi sets may be susceptible to dimmer interference. See Table 4-5.

Figure 4-15 shows several other types of dimmer arrangements which might be useful in various situations.

In Fig. 4-16 is shown the principle of recessed lighting. This may be used to good effect in playrooms, in dining rooms, or other places where light is desired but where you do not want to see the fixture. The size of the light, or the number of these kinds of units used, will depend upon the primary use of the room to be lighted. When placed at an angle, they can be used to highlight walls or objects of interest. There are some other suggestions with regard to common lighting situations shown in Figs. 4-17 through 4-21. These illustrations are courtesy of G.E. A note to them at Nela Park, Cleveland, Ohio, will bring you even more details on various lighting schemes and ideas for your home or office.

In Fig. 4-21 are shown some types of lamps. Equally important are the concepts pertaining to the shades. Some general information on shades is as follows:

Shade size and shape—Open top recommended. Minimum dimensions: 16 in. bottom. Note: Shallower shades need louver or shield in top. Deeper shades need extra-long harp or recessed top fitting.

Shade material for reading lamps:
a. Moderately translucent preferred to opaque.

INSTALLATION CIRCUIT
DIAGRAM FOR 3-WAY CONTROL
INCANDESCENT

TYPICAL INSTALLATION
INCANDESCENT AND FLUORESCENT*

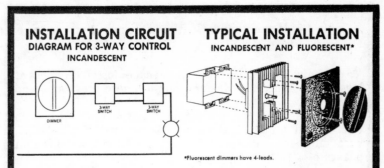

*Fluorescent dimmers have 4-leads.

Fig. 4-13. A typical dimmer circuit installation for incandescent along or installation (**not circuit**) for incandescent and fluorescent light.

SYSTEM WIRING DIAGRAM

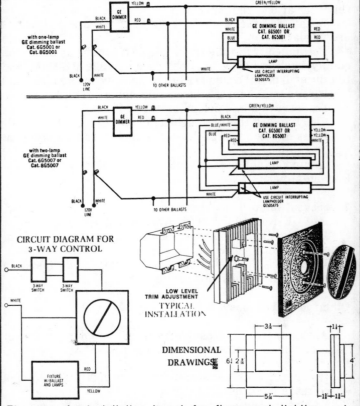

Fig. 4-14. An installation layout for fluorescent lighting, using a "Specifier" dimmer control. (G.E.)

Table 4-5. Recommendations for Dimmer Performance

To minimize radio frequency interference: Install dimmers at least 12 ft from intercom and public address systems and radios.

Wire intercom, P.A. systems or radios on separate branch circuits and in individual metal EMT or conduit. If possible, wire dimmers and communications systems on different phases of a 3-phase system.

Keep intercom leads at least 6 ft away from ac lines feeding dimmers.

When intercom wires and ac lines feeding dimmers must cross, be sure to cross at right angles.

Wire P.A. speakers with shielded conductors—ground shielding at one end only.

Install a Cornell-Dubilier IF18 filter when P.A. systems are in place.

Permanently installed microphone cable should be of high quality, such as Belden or Alpha coaxial cable RG-55/M and run through metal conduit. Keep dimmer connected load and line wires a minimum of 3 ft from microphone cables when run in parallel, whether shielded or not.

Check to see that microphone cables are tight.

Cross microphones cables with dimmer load wires only at right angles.

Ground amplifier case. Also, ground microphone cable at P.A. amplifier.

To prevent lamp hum:

Use GE No. 200AX lamps when possible.

Lighting fixtures should be nonparabolic to prevent focusing of lamp noise.

When GE No. 200AX lamps are not used, install a debuzzing coil on load side of dimmer.

Other considerations:

All dimmers are sensitive to transient voltages. Never "spark" wires together to check for power. Always use a circuit tester.

It is normal for dimmer faceplate to feel warm. Mount where there is adequate air flow.

When mounting dimmer, heat sink should **not** be recessed.

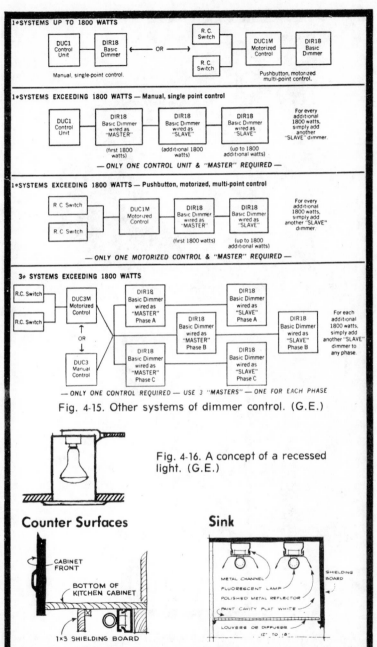

Fig. 4-15. Other systems of dimmer control. (G.E.)

Fig. 4-16. A concept of a recessed light. (G.E.)

Counter Surfaces

Sink

Fig. 4-17. Installation of fluorescent lights at the sink and under the counter. (G.E.)

15" to 18"

9" to 12"

Fig. 4-18. An effective lighting system for the student (G.E.)

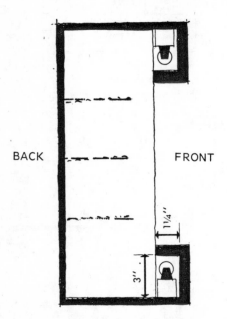

BACK

FRONT

1¼"

3"

Fig. 4-19. How to illuminate a shelving arrangement. Place Slim-Line deluxe warm-white fluorescents in front corners behind the trim as shown. (G.E.)

b. Moderate-transmission white vinyl or parchment, laminated to fabric—or white-lined fabric, if fairly dense.

c. All shades in room similar in brightness.

d. Dense or opaque shades only when walls are very dark.

Shade material for makeup lamps—Use highly translucent material (thin plastic, fiber glass or silk). Never use fabric laminates.

Shade color—Avoid strong or dark color on outside unless material is opaque. Inside always white or near white, never shiny. Colored linings tint the light. Foil linings reflect the light meagerly and with harshness.

One type of fixture which will provide light without glare over a work area, or a play area, is the twin mount light fixture made by Owens-Corning Fiberglas Corp. Its installation is shown in Fig. 4-22. There are some photographs of a finished installation in Figs. 4-23 and 4-24.

INSTALLING FLUORESCENT FIXTURES

So many times we are asked about the actual installation of the fluorescent light that we include next a discussion of the

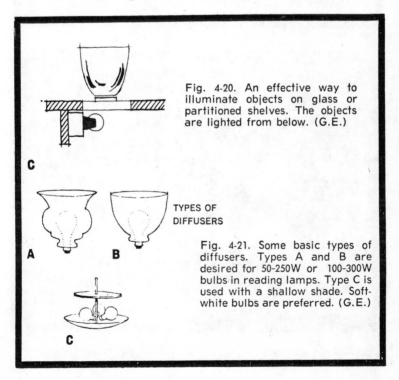

Fig. 4-20. An effective way to illuminate objects on glass or partitioned shelves. The objects are lighted from below. (G.E.)

TYPES OF DIFFUSERS

Fig. 4-21. Some basic types of diffusers. Types A and B are desired for 50-250W or 100-300W bulbs in reading lamps. Type C is used with a shallow shade. Softwhite bulbs are preferred. (G.E.)

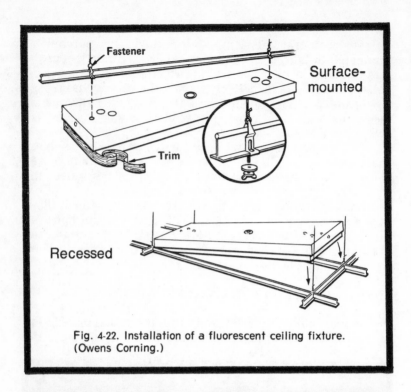

Fastener

Surface-mounted

Trim

Recessed

Fig. 4-22. Installation of a fluorescent ceiling fixture.
(Owens Corning.)

Fig. 4-23. A ceiling mounted fluorescent fixture. (Owens Corning.)

wiring for structural lighting of this type of fixture. The component parts of the fixture are shown in Fig. 4-25, and a special type of mounting, the **side mounting channel,** is shown in Fig. 4-26.

Since structural lighting is part of the permanent structure of the home, the lighting elements should be permanently wired and switched. It is important that the **wiring outlet** be carefully located on the wiring plan so that it will be covered by the fluorescent channel in the final installation. All fluorescent lamps perform better when used on grounded wiring systems and in metal wiring channels.

Some channels are available equipped with special ballasts that provide quick starting without a separate starter in the circuit. In the 15 and 20W sizes, "trigger start" ballasts will give quick starting for regular fluorescent lamps. In the 30 and 40W sizes, quick starting can be obtained with "rapid start" ballasts.

In every case where lamps are to be used in continuous rows, it is important to select channels that have their sockets mounted at the extreme ends of the fixture. These may be butted back-to-back to provide a continuous smooth line of light with minimum socket shadows.

There are numerous occasions when the flexibility afforded by dimming is effective in adjusting the lighting to fit

Fig. 4-24. A method of making a recessed installation, panel type, fluorescent fixture. (Owens Corning.)

Fig. 4-25. The parts of a fluorescent fixture:

1. UL listed channel maintains socket spacing, safely contains electrical parts, aids in lamp starting.

2. Ballast is required to stabilize lamp operation. Desirable features are "certified" label, "high power factor" and "A" or "B" sound rating.

3. Starter is required with conventional ballasts and lamps. Specify "certified" starters—FS-2 for 20W lamps, FS-4 for 40W lamps. (G.E.)

the mood. A number of practical dimmers are available for the control of fluorescent lamps in residential installations. They all will fit into 2x4 stud walls and are easy to install and operate.

Dimming fluorescent lamps is easy and practical, but there are a few points to remember:

1. As a general rule, fluorescent dimmers require a three wire cable connection from the dimmer to the fluorescent channels.

2. Special dimming ballasts are required for each channel and lamp.

3. Dimming systems must be grounded.

4. Dimming systems will operate the 30 or 40W rapid start lamps, but will not operate them satisfactorily together on the same circuit.

WALL BRACKETS

The wall bracket is probably the single, most useful structural lighting device in the home. It can be used in any room of the house. Basically, there are two kinds of wall

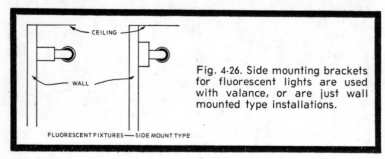

Fig. 4-26. Side mounting brackets for fluorescent lights are used with valance, or are just wall mounted type installations.

3" MIN. 2" MIN.

CHANNEL MOUNTED EVEN WITH TOP OF FACEBOARD

INSIDE FLAT WHITE

Fig. 4-27. The method of mounting a light behind a wall faceboard. Note the dimensions. (G.E.)

brackets; the construction differs depending on whether the bracket is to be used high on the wall for general lighting, or lower on the wall for specific task lighting.

A high wall bracket is really a valance without a window. It is used as a source of general lighting for a room. Quite often it will be used to balance the illumination from a matching valance at an opposite window.

Many of the same dimensions and construction techniques must be observed as when installing a valance. The fluorescent lamp and channel must be located as high up behind the shielding board as possible so the light will spread evenly and far out over the ceiling. The fluorescent tube should be at least 3 in. out from the wall to provide a smooth distribution of light over the wall. This reduces the chance of hot streaks of brightness above and below the shielding board.

A minimum of 10 in. between the top of the shielding board and the ceiling is recommended so no light is trapped above the bracket. See Figs. 4-27 and 4-28.

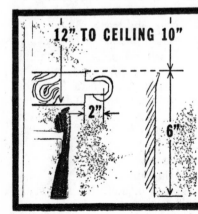

12" TO CEILING 10"

2"

6"

Fig. 4-28. The location of the valance type fixture may be critical. Note the distances. (G.E.)

The inside of the faceboard should also be painted flat white. The only exception to this would be the bottom 1½ to 2 in. on a very deep valance. It is possible that this bottom strip might be seen from outside the window and therefore can be painted to match room decoration. If the back of the faceboard is not flat white, the valance will trap the light produced, and the resultant color of light in the room may be distorted.

VALANCES AND DRAPERY

The most critical dimension of a valance is the spacing between the fluorescent lamp and the front surface of the drapery material as shown in Fig. 4-28. There must be at least 2 in. between the center of the lamp and the front face of the drapery material. This insures that the draperies will be more uniformly lighted from top to bottom. Draperies hung near the tops of their pleats hang straighter, causing less interference with the light. To get the proper spacing it is usually necessary to allow 3 in. between center of the lamp and the drapery track. Often, this requires mounting the fluorescent channel out from the wall by means of wooden blocking or metal straps.

In normal drapery installation conditions, the inside of the faceboard should be located about 6½ in. from the wall. If bulky draperies, or a double track is used, the faceboard might have to be extended to a distance 8 in. or more from the wall.

LIGHTED CORNICE

The lighted cornice is positioned on the ceiling at the junction between the wall and ceiling. All of its light is directed downward to light the wall surface below. For this reason, the lighting effect produced is dramatic. It emphasizes wall textures, wall coverings, and will light pictures and other wall hangings. Also, because the wall is emphasized, the cornice gives an impression of greater ceiling height. Cornices, therefore, are ideally suited for low-ceiling rooms such as basement recreation rooms. The cornice is about the simplest of all structural lighting elements to build. There are, however, a few points to note; see Fig. 4-29. These key points should be followed when installing a cornice:

1. There should be 2 (preferably 3) in. between the center of the fluorescent lamp and the surface to be lighted.
2. Paint the inside of the faceboard flat white.
3. The channel should be positioned as close to the faceboard as possible. The faceboard should be at least 6 in. deep.

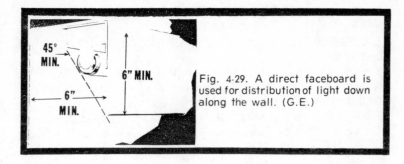

Fig. 4-29. A direct faceboard is used for distribution of light down along the wall. (G.E.)

45° MIN.

6" MIN.

6" MIN.

COVES

Coves are particularly suited to rooms with two ceiling levels. In these applications they should be placed right at the line where a flat, low-ceilinged area breaks away to a higher-ceilinged space. The upward light emphasizes this change of level and is very effective in rooms with slanted or cathedral ceilings.

The lighting efficiency of coves is low in comparison with valances and wall brackets. Because of this, more lamp lumens are usually needed. To attain a general lighting level of from 5 to 10 foot-candles in a living space, coves should be designed for 45 to 60 lumens per sq ft of floor area. Coves (which are usually mounted high on the wall) direct all of their light upward to the ceiling where it is, in turn, reflected back into the room. The cove is known as a source of "indirect lighting." The illumination effect produced by cove lighting is soft, uniform and comfortable. Since there is no light directed downward into the room from a cove, however, the resulting lighting effect is relatively flat and lifeless. For this reason, cove lighting should be supplemented by other lamps and lighting fixtures to give the room interest and provide lighting for seeing tasks.

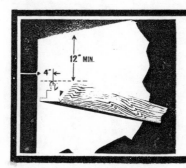

12" MIN.

4"

Fig. 4-30. A method of adding cove lighting to a room . (G.E.)

For good cove lighting a few basic rules must be followed; see Fig. 4-30.

1. Cove lighting should only be used with white or near-white ceilings.

2. Keep the cove as far down from the ceiling as possible for a wide distribution of light. There should be a minimum of 12 in. from the top of the shielding board the ceiling.

3. Place the lamp at least 4 to 4½ in. out from the wall.

4. Paint the inside flat white. Butt lamp sockets back-to-back.

Interior Decorating
With Light

How often, when visiting friends, have you thought to yourself, "How pleasant and relaxing their home was! How nicely the colors blended. What wonderful taste in decor she has!" You were so conscious of the dramatic and interesting effects of everything in the home, but perhaps not so conscious that it was the lighting which brought out all the best features.

Sometimes the reverse happens. When visiting people you come away with the weariness and irritability which only undue eye strain could cause. "The home was nice—**but** the **glare** was awful! They need to do **something** about their lighting!"

Home lighting can make a big difference, not only in our friends' opinions, but also in our own peace of mind, comfort, and state of relaxation. Good practices can inspire warmth and companionship, stimulate joy and happiness—bad lighting can induce melancholia and despair. Everyone needs to examine the lighting in his home and then to make it a project to obtain the kind of lighting which produces the atmosphere he most cherishes.

We want to obtain now a combined view of the principles of illumination from the **decorators'** and the **architects'** viewpoint. Sometimes these are called the engineers' and the interior decorators' viewpoints. They are both sciences, one concerned with the esthetic beauty of lighting, and the other, the engineering viewpoint, with the utilitarian aspects of lighting. Let us begin with a review of colors and reflectance values of colors.

PRINCIPLES OF LIGHTING DECOR

All decorating is a part of interior lighting design—for every surface reflects some of the light it receives. Light can be absorbed and even wasted by dark surfaces, or it can be reflected by lighter surfaces and utilized as useful illumination. Remember that this reflected light will always

Table 5-1. Recommended Reflectances for Major Surfaces

	REFLECTANCE PERCENTAGES	
	MINIMUM	MAXIMUM
Ceilings		
Pale color tints	60	90
Walls		
Medium shades of color	35	60
Floors		
Carpeting, tiles, woods	15	35
For extensive down-lighting installations	35	60

be tinted with the color of the reflecting surface. It is this principle of reflected light that is often responsible for making the color of painted walls appear more intense than the color of a small paint chip.

Light makes colors live. In low levels of illumination, colors are grayed and lifeless. As illumination increases, they become more vibrant and alive—even pale tints and deep wood tones. Table 5-1 is a simple guide of recommended reflectances for major surfaces.

Room Unity

When planning the lighting for a room, you must consider the unity of the room. The floor plan and the furniture must be considered; the kinds of materials, the colors of the walls, the effect to be achieved in the room. You must also consider the **type** of lighting, that is, will it be best to have single lights, groups of lights, or supplemental lighting as is usually required for period-type fixtures. You must consider the **kind** of lighting, that is, whether fluorescent or incandescent, and consider the candlepower levels or lumens which are appropriate to the situation.

When considering the **size** of the rooms to be lighted, remember that in a large room **less** light is absorbed by the walls, but due to the area, more lights may be required. Fixtures must be considered not only for their ability to produce the desired lighting effect (illumination plus beauty or interest) when lighted at night but also their effect on the room during the day when they form an important part of the space in the room. In small rooms, the height of the light may be very important. More light will be reflected (or perhaps should be) from the walls. Undesirable reflections may take place and "hot light spots" may exist. Multiple lighting may also be called for to achieve the desired overall **level** of illumination.

In a general situation where good visibility is a prime requirement, it is usually good practice to have the lights as high as possible. It is also good practice to have symmetry in the arrangement of the overhead lighting fixtures.

When planning an arrangement of lights, there are several "decor" considerations. If, for example, the light is so placed that it "grazes" a wall surface, such as shown in Figs. 5-1 and 5-2, it will enhance the texture of the wall, and produce a dramatic effect. It will produce a striking effect in lights and shadows due to the surface itself. If, however, you use a "frontal" type of light playing on the wall surface, it will tend to reinforce the concept of flatness, and will minimize any blemishes or projections the wall may have. It will enhance the idea of room size.

Of course, you may use a kind of "isolated highlighting" on walls or objects to produce dramatic special effects such as shown in Fig. 5-3. This kind of lighting can introduce color variations, or enhance color settings if the light is the same color as the setting, or produce contrasting moods when dif-

Fig. 5-1. A grazing light brings out the wall texture. (G.E.)

Fig. 5-2. A combination of cove and valance lighting over the fireplace adds height to the room and gives dramatic interest to the brick wall. (G.E.)

ferent colors mix. It may be used to produce "sparkle" which is the kind of unshielded lighting you have on Christmas trees small uncolored bulbs and other such objects. Sparkle may be desired in candelabra to enhance their own graceful design, and yet not detract from the beauty of the fixture itself. Isolated lighting or "small area" lighting, that is, using a directed light in a small space on the wall or in the room, can be used to produce different lighting levels. These, in turn, produce by themselves a dramatic and contrasting effect in the home. They create beauty and interest scenes. Examine Figs. 5-3 and 5-4 and add interest to scenes.

Lighting Effects

You may use light to expand or contract actual space by creating "light perspectives," as in Figs. 5-5 and 5-6. For example, to increase the height in a given room, use vertical panel lighting and fixtures. Radiate the light up from the center of the walls so the light becomes dimmer toward the ceiling and thus creates the illusion of distance. Use long chains on fixtures to give an added impression of height. Conversely, to bring a high ceiling down, light up the area near the ceiling brightly and let the shadows or subdued light diminish in the floor direction. Use short chains on fixtures and **do not** use vertical panels of light. (A panel of light may be a fluorescent fixture mouted vertically with a plastic covering so the light itself is not revealed.)

When you desire to **widen** a room, place the panels **horizontally** in the room and repeat the patterns horizontally. Use small areas of wall which are lighted in rectangles and use many of them. This same procedure, the parallel lines of light—the rectangles which are narrow in height and long—can be used to increase the length effect of a room. Of course, you can always highlight interesting objects in the room such

Fig. 5-3. A highlight effect on a painting produces interest. (G.E.)

Fig. 5-4. For formal occasions the candelabra with its "sparkle" creates a pleasant and lovely atmosphere. (G.E.)

as planters, sculptures, paintings, etc. with appropriate colors and intensities which result in beauty and drama. See Fig. 5-7.

COLOR AND LIGHT LEVEL

Speaking of colors, there are some nice effects with which you should become familiar. For example, my wife brought home a long cylindrical fixture which was sculptured and gold tinted. It was designed to hang from a gold chain and we had a proper place for it in the living room near a large painting which also had the yellow, gold, reddish colors predominating. When we hung it and turned it on (with a low level of white light—a 25 watt bulb), it was horrible! You have already guessed the answer. We replaced the white with a yellow bulb of the same intensity. All of the immediate color was enhanced and the room became bright and cheerful and warm and

Fig. 5-5. Light filtered through louvered panels provides a soft intimate atmosphere. (G.E.)

Fig. 5-6. Overhead lighting gives height and brightness to the setting of Fig. 5-5. Note how a new atmosphere can be created by using an alternate lighting system. (G.E.)

pleasant! So you too must be aware of this "color effect." Note that red, pink, or yellowish light in low levels tends to enhance the human complexion much as the hue or tint control does on your color TV. At the same time, blues, greens, magenta, etc., will give a greenish cast to complexions. The effect of this latter kind of light is to produce solemnity, calmness, and spiritual reverence. Sometimes one desires greenish lights to enhance plant areas or rooms which are decorated in greens. This means, of course, that you should use the color which enhances the color scheme of the room in which the light is to be used. Sometimes when one has a room lighted in gold, as was ours, a small gold-color spotlight may be added to shine through the gold haze of the fixture lamp. A startling effect is thus produced which heightens the effect of the already visible coloring. It's worth experimenting!

Fig. 5-7. Highlighting for effect. A single spotlight brings out the beauty of the table and creates a subdued atmosphere for dining. (G.E.)

Fig. 5-8. The brightness without glare in this pleasant living room gives a warm and happy feeling to everyone.

The level of illumination has a psychological effect on people. For a gay and happy atmosphere, as in Figs. 5-8 and 5-9, a high level of illumination is desirable and there should be changes in levels from room to room or area to area and a change in the color schemes. Warm colors, of course, should be used. Reminds me of a novelist friend of mine who said "you cannot create a state of happiness unless you have a contrast...spend a moment considering depression and then happiness!" Lighting can create the same kind of emotion—a happy area is much happier with its warm brightness if one enters it from a lower level of subdued lighting!

Now let us consider still more definitions of lighting, Table 5-2, relating to other than lamps or pendant fixtures which you might use in your home. It well might be that you would want to make such additions to your current lighting, or make changes in your present lighting system. Some decorators now believe that a two- or three-fold ability to change lighting in a given room to change moods and tones will immeasurably increase your living atmosphere. You can change from one kind of lighting to another simply by turning switches on or off.

A kind of lighting which is very suitable for a baby's room, giving bright illumination, yet no direct glare, is shown in Fig. 5-10. Note that it casts the light both up and down on the wall.

The intensity can be varied through the use of dimmers and fixed fluorescent tubes so it is soft and pleasant. Valance lighting is shown in Fig. 5-11.

A luminous cornice is a panel of light (long fluorescent tube) which is mounted at the intersection of a wall and the ceiling. See Fig. 5-12, over the painting. Flush spotlights are used in this figure to highlight the curtains and provide a soft room-light effect.

Let us consider some **bad** things about lighting which you need to avoid if your home is to be tasteful and comfortable from a lighting standpoint. The first DON'T is: DON'T HAVE

Fig. 5-9. Good visibility creates a feeling of well-being and adds space to the rooms. (G.E.)

Table 5-2. Some Definitions of Lighting

Luminaire	A complete lighting unit of lamps, reflector protective cover and connections.
Regressed	Opening of the luminaire above the ceiling.
Flush or recessed	Opening of the luminaire on a level with ceiling.
Matte Surface	A surface which produces a diffuse reflection.
Cove light	A light shielded by a ledge, or a horizontally recessed light in ceilings or walls.
Cornice light	A light source shielded by a panel parallel to the wall, attached to the ceiling. It distributes light over the wall.
Coffer	A recessed panel or dome in the ceiling.
Troffer	A long recessed lighting unit installed with its opening flush with the ceiling.
Valance	A longitudinally shielded member mounted across the top of a window or wall, usually parallel to the wall. It conceals light and gives both up and down illumination.
Louver	A series of baffles to shield the light source from certain angles of view, and which will absorb unwanted light.
Lunette	A light which is small and unobtrusive which is placed high in a wall opposite to the object upon which the light is focused.

GLARE. This may come from a light itself which is a hot spot and overly bright, or it may come from the reflection of light on glossy surfaces of any kind. All lighting in the visual field, the normal viewing area, should not be overly bright, but always at a comfortable light level. A lamp which is seen as a bright object against a dark wall is bad. If you need a bright lamp, then the wall contrast should be small and the wall should be bright also, as in Fig. 5-13. If there's a dark wall such as a paneled den, then the lamps should have shades which blend with the walls and produce a subdued light which will not be reflected and will not produce glare from the polished walls. Often, an additional kind of lighting may be necessary such as from coves or some kind of recessed panel lighting in walls or ceiling.

There is another consideration: When you are seated in a room, there are considered to be three areas to your vision—the **near vision** on the task (book, etc.), the **intermediate vision** on nearby objects, and the **far vision** areas which are around, and beyond near and intermediate. When we plan the light for a room, there should not be deep shadows or large contrast seen if one looks up from the task area. The change should be gradual and even, Fig. 5-14, whether going to a higher or a lower level of lighting. As you know, your eyes will have to adjust for the different levels of light. If there are strong contrasting areas in a room, as in Fig. 5-15, and you look at them often, then your eyes work hard and you can experience eye fatigue, possible headaches, and irritation without realizing the cause. **Pleasant and comfortable lighting** over all the areas will save your nerves and enhance your pleasure.

FIXTURE TYPES

Let's examine some basic examples of the kinds of fixtures we want in the home or office. Figure 5-16 defines five different types.

Fig. 5-10. In the baby's room, a good light without glare is essential. This light should also be dimmer-controlled to reduce the levels of illumination for sleep periods. (G.E.)

Fig. 5-11. Valance lighting in a small sitting room minimizes the difference between day and night since the light comes from the window in a wide pattern.

Fig. 5-12. A combination lighting system. Note the spotlights on the curtains giving a varied density of light pattern and the method of illuminating the wall and painting with **cornice** lighting. (G.E.)

Direct Lighting

This type fixture has a reflector so arranged that it beams the light downward on the task area. It may produce shadows elsewhere if it is not high enough for general illumination. It can give a bright, high level of light for the task area if mounted overhead. As in Fig. 5-17, it will produce shadows on the task area.

Semidirect Lighting

This fixture has some radiation of light upward, but it is small. If this fixture is mounted too close to the ceiling it produces a pool of light which can become a "hot spot" as far

Fig. 5-13. A student study desk is lighted for maximum study capability and a pleasant atmosphere. Note that the walls do not produce a large contrast. (G.E.)

Fig. 5-14. The book case is highlighted for interest. Note the level and the type of reading lamp used. A second type of illumination may be obtained by turning on the recessed ceiling panels. (G.E.)

Fig. 5-15. This illumination scheme is not recommended. There is too much contrast and change in level from light to dark. The lamp seems to create an intense bright spotlight effect, almost a glare.

ICI* Light Distribution	Type of Luminaires	Type of Mounting
Direct 90% to 100% Downward	Recessed	In Ceiling
	Individual Unit Continuous Row	On Ceiling
	Individual Unit Continuous Row	On Ceiling
	Luminous Architectural Elements	On Ceiling, Side Walls, or Columns
	Decorative Ornamental	On Ceilings or Walls
	Large-Area Low-Brightness	Recessed in Ceiling
	Diffuser Ceilings Louver Ceilings	From Ceilings
Semi-Direct 60% to 90% Downward	Individual Unit Continuous Row	On Ceiling
	Individual Unit Continuous Row	
	Luminous Cornice	On Wall near Ceiling
General Diffuse 40% to 60% Downward	Individual Unit	On Ceiling
	Continuous Row	
Semi-Indirect 60% to 90% Upward	Individual Unit Continuous Row	On Ceiling
	Luminous Cove Wall Bracket With Downward Component	On Wall
	Luminous Ornamental	On Ceiling
Indirect 90% to 100% Upward	Individual Unit Continuous Row	On Ceiling
	Wall Urns Cove Pedestal Ceiling Soffit	On Wall On Wall or In Ceiling Floor or Counter Top In Ceiling

Fig. 5-16. The accepted definitions of the types of lighting systems. The patterns to be obtained are shown on the left.

Fig. 5-17. This illumination system can be improved. Adding a recessed fixture over the sink area prevents shadows on the tasks performed there. (G.E.)

Fig. 5-18. An atmosphere for comfort and efficiency. She will always look her best as she makes up in this room. (G.E.)

as vision is concerned. If it is mounted a proper distance from the ceiling it is helpful in eliminating shadows and gives good general lighting.

Diffuse Lighting

The fixture is covered with a frosted cover of plastic or glass which is designed so it does not produce glare. The illumination is soft and general. It is not used, generally, in areas where the tasks require a high level of illumination. It is most used in the intimate situations (as exemplified in Fig. 5-18) and where a confidential and picturesque atmosphere is desired.

Fig. 5-19. A bright bath area. No shadows will conflict with shaving using this type of mirror light. The general level of illumination comes from the ceiling panels. (G.E.)

Direct-Indirect

This fixture has some glare but is okay generally. It gives a room light level and a direct task level which is good. Many lamps are designed to produce, by a three-way switch, the **indirect,** the **diffuse,**or the **direct-indirect** type of lighting so you can change lighting as you desire for different atmospheres and occasions. A **semi-indirect**lighting fixture has most of the light directed upward and a lower level directed downward on the task.

Indirect Lighting

The light from this fixture is all **reflected from the ceiling** and a uniform ceiling illumination is needed to obtain a smooth and even lighting over the room and eliminate "hot spots" of light from the fixtures which throw the light upward. "Hots spots" produce glare and focus attention to areas where it is not desired to have the attention focused.

One more example of **diffuse** type of lighting, creating a most pleasant atmosphere, is shown in Fig. 5-19. You may want to consider this type of lighting for some rooms in your home.

Exterior Wiring and "Lightscaping"

Chapter 6

Beautiful homes and offices are beautiful by day or by night. Safe homes or offices are made safer at night with light. Light is the magic touch of the artist with the practicality of the most severe realist. We now consider the use of light for exterior locations around the home or office (Fig. 6-1), and the wiring methods which add liveability and safety to the areas around and outside the building by adding the flexibility and convenience of electricity to these areas.

It was FUN spelled with capital letters in my own case when I began preparing this book, to examine and use some of the many ideas which developed from the necessary research. In the front yard, for example, we had long been content with a small white spotlight illuminating a portion of the wall and a small light at the entranceway, low and almost hidden by shrubbery, to give some light to the pathway to the front door. How inadequate this turned out to be, and how fascinating the changes which we made to give drama, color, texture, light and safety to the exterior areas around the house. The single white spotlight became a pair of blue-green lights which highlighted pyracantha, blue spruce and evergreens. The low-shrubbery path light became higher and amber to highlight the walkway. A new, small 7½W spotlight was placed high in the entrance alcove, hidden by a structural beam, and this proved to be exactly right to produce just a soft glow on the walk and entranceway. Yard lights on attractive poles illuminate the back yard area and make it safe for us, rather than a good hiding place for whatever creatures roam at night. Most of all was the fun in installation, and trials and imaginative processes which we employed and are now continuing to employ to make our home beautiful at night. One example of entranceway lighting is shown in Fig. 6-2.

You **can** get carried away. You become conscious of lighting effects of the neighbors and friends, and you make tours through town to see just how others are doing it. It's a fascinating hobby, and a most practical one also. Just as with the **interior** lighting you become conscious of all the homes and

Fig. 6-1. One method of house illumination. Note the highlight for the small tree, and the corner of the house. (G.E.)

offices and stores and restaurants you visit, and you notice and inspect the multivaried lighting effects which are used by the experts. Examine some displays in store windows and you'll see how color and intensity and placement of lighting is used to the ultimate advantage. You'll become an expert yourself as you try and use various effects.

Once you have the "bug"—permit me to be trivial and say "lighting bug"—then you'll want to know **how** to do it. How to install the wiring and fixtures and things; and you'll want to know what is available. Loran, Inc. of Redlands, Cal. has provided much of the basic information herein on exterior lighting systems. Loran makes components for such light-scaping systems. The systems it makes and sells, under its Nightscaping trademark, are **low voltage** types; and so you do not run the dangers which you might have with the higher voltages. We shall discuss both the low voltage systems (12V) and the extensions of the regular house currents and voltages (120V) for the various ideas to be examined.

EXTERIOR LIGHTING CONCEPT

Let's examine the exterior lighting concept. First of all we answer the basic question of "Why do it?" When you employ exterior lighting or "lightscaping":

a. It assures you of more safety in the dark areas of your property.

b. It illuminates the beauty of your gardens and grounds.

c. It creates dramatic and interesting silhouettes, shadows, and highlights.

d. It can influence the mood of owner and guests.

e. It makes evenings outdoors more pleasant and liveable.

f. It discourages uninvited visitors and prowlers.

g. It gives illusion of greater size and depth to property.

Fig. 6-2. A well lighted entranceway seems to invite guests and denotes a feeling of bright happiness. (Loran, Inc.)

Fig. 6-3. Creating dramatic interest using shrubbery shadows. (Loran, Inc.)

h. It makes warmer and brighter welcome for guests.

i. It adds to both the esthetic and investment value of the property.

j. It makes for a more pleasant way of living.

TYPES AND EFFECTS

Next we consider some important definition.

Uplighting—The fixture itself is at a low level and will beam its light upward. This can create dramatic shadows on a wall, Fig. 6-3, by placing the light directly below selected plants and bushes, or other foliage, and directing the light upward through them as in Fig. 6-4.

Downlighting—Here the fixture is high, above the plants, bushes, etc. The light source should be hidden from view, and should always beam the light straight downward and not off to any angle. This way maximum effect is achieved in uplighting.

Highlighting—The fixture is so placed that it directs its beam on some object or focal point in the yard, or on the house. Fig. 6-5 is a good example of this kind of illumination. Such objects

Fig. 6-4. The skeleton of the tree itself becomes a vital point of interest when illuminated from below. (Loran, Inc.)

Fig. 6-5. Highlighting in the garden area provides beauty in a pleasant setting. (G.E.)

may be a fountain or a pool or some art object such as a statue or perhaps a particularly beautiful plant or other living thing which has beauty when lighted, or which will give a dramatic effect with its lights and shadows. When highlighting, the intensity is **increased** and the light source is placed so as to give the desired effect on the object. The light source itself should be concealed as much as possible.

Fig. 6-6. A well lighted home discourages prowlers. (G.E.)

Floodlighting—When you illuminate the house as a unit then a gentle light is desired. It should be soft and complete (Figs. 6-6, 6-7). It may vary in color and direction and may vary in intensity so as to create effects, but do not try to illuminate harshly as is done in a parking lot or around a store front.

Cross lighting—When light comes from two different sources and plays on an object or an area, this is called cross lighting. Such light may cross in itself and illuminate individually the far reaches of the beams. Sometimes the illumination is on the front and back, and in this case both beams play on the object but from different directions. The sources are usually at ground levels and it is customary to use different intensities of light from the various sources. We have said **two** beams, but there is no reason why three or more might not be used and the use of color and intensity be combined to give the maximum kind of dramatic, interesting or beautiful effect.

Underwater lighting—How beautiful it is in a garden pool or a swimming pool. When the fixtures are underwater (Fig. 6-8), a special kind of fixture is required. Also a special wiring procedure is required to insure safety and nondeterioration of the equipment. (Loran, Inc., makes equipment for such locations.) When water has ripples there are varied effects as

Fig. 6-7. Highlighting for a dramatic effect. Concealed low level spotlights provide light and shadows. (Loran, Inc.)

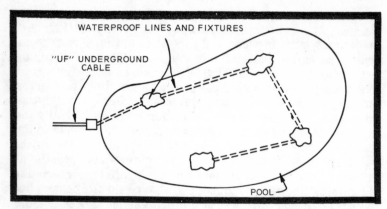

Fig. 6-8. A simple method of adding underwater lighting to your swimming pool. Of course, the proper type of electric cables and fixtures must be used.

light passes through it; and thus fountains or small waterfalls add beauty and drama in a yard or garden, when properly illuminated from behind or inside. The light sources may be colored for effect, and the intensity varied as desired by use of low or high wattage bulbs. The light sources should be concealed either as objects of the structure itself, or hidden in alcoves or troughs.

Fig. 6-9. Light is provided for the area in which guests will gather, and the light will be proper for the kind of games to be played, or to create the atmosphere desired. Note the auxiliary lighting on the fence and over the lawn seat. (G.E.)

Recreational lighting—When illuminating a game area, Fig. 6-9, one must be careful and think over the kinds of games being played. The lights must **never** interfere with the player's ability to demonstrate his or her skill. The lights must be bright, but removed from the normal field of vision. Glare must be controlled to the maximum extent possible and eliminated if possible. The lighting equipment itself must not interfere with the play or present a safety hazard. The intensity of illumination in foot-candles for this kind of area was discussed in Chapter 4. You may want to refer back to that discussion, and the tables therein.

Antiprowler lighting—A low ground light, Fig. 6-10, will discourage prowlers more effectively than a harsh spotlight. There should be switches on this kind of lighting, and these may be the kind which turn off and on **automatically** with the beginning of sunlight in the morning and termination of sunlight at night, or they may be controlled by regular house switches. We will discuss this in more detail later in this chapter.

Fig. 6-10. Illumination of areas with small, inconspicuous lamps. (Loran, Inc.)

Fig. 6-11. How pleasant the outdoors can be! Proper uplighting creates a most comfortable distant view. (G.E.)

Safety lighting—This type of illumination is strictly for utilitarian purposes and is used to prevent accidents and damage suits. It is any type of lighting which will provide proper illumination for walks, pathways, paths, steps, and dark and hazardous areas around and even **in** the house and other buildings. One example in our own home: We had a light in a small patio which provided illumination to the front door. It was located behind a small shrub which, of course, grew. The illumination pattern changed so that above about 4 ft one could see easily, but the **pathway** with its attendant dangers due to curb, etc. became dark and indistinct. The solution was to add a small light placed high to provide **downlighting** over the entire entranceway, but keeping the illumination level very soft and inconspicuous. This was a necessary addition to the esthetic lighting effect of the low level shaded lamp in order to obtain the required **safety** on the pathway.

Vista lighting—This is used for a distant section or area of the garden, Fig. 6-11, or grounds or where a clear view of the far countryside, the mountains, or the city is desired. Lighting in this situation is accomplished with all light sources subdued and mounted below knee level so that the distant view will not be blocked by excessive lighting glare. No fixtures are visible.

Backlighting—In this process, plants or other objects are placed in silhouette. This is one of the most dramatic lighting effects possible to achieve. One should experiment with a movable light source or use several light sources placed at different angles and with different size bulbs (to obtain different intensities) and with different colors to get the most pleasing and dramatic effect possible. This is a fun project for the whole family.

Mirror lighting—When you have water available in small ponds or swimming pools, you might desire to have a **double** illuminating effect, Fig. 6-12. Use underwater lighting, and then plan a second scheme which can be used as an alternate, which will utilize the reflective characteristics of the water. In this latter case, you plan an arrangement of plants or statuary, or both, **behind** the area, in the line of view, which can be brought into illumination effectively. Then the illumination sources are arranged so that the light is reflected off the pool of water onto the planned area. The effect is magnificent! It becomes an equivalent moonlit expanse of beauty. If you have a small lake or natural pool on the property, it is ideal for this kind of lighting effect. Try it!

Skeleton lighting—What a name! It implies what it says, though. You try to highlight grotesque shapes of tree trunks and branches, especially as they appear in the winter season.

Fig. 6-12. The pool adds its reflections to increase this bright scene. (G.E.)

A blending of the barren and desolate and the evergreen makes for a magical scene. Look at some of the fine paintings of just such subjects! Lighting can highlight a living picture and scene for you effectively. You will want to try this type of lightscaping also.

PLANNING AND WIRING

Now let us discuss some procedures and practical aspects of the projects. First of all there are some general rules to be observed:

1. Experiment with the fixture placements **at night**. Move them around and test them before making a final or semifinal installation.

2. Always leave at least 12 in. of slack wire at each ground-mounted light location so that you can adjust the light as plants and shrubs grow.

3. Be sure to hang all lights **above** the lowest limb of a tree if you use tree limbs for light supports. This will cause attractive shadows on the ground and eliminate harsh direct glare.

4. Be sure to protect all lines with fuses of proper voltage and amperage rating.

5. Be careful that you do not overwork a few light sources. Use as many as necessary to give the desired effect. Be generous!

6. Do not **overlight** some object you wish to bring into focus. Overlighting is one of the most common mistakes in first attempts at illumination control.

7. Never install fixtures closer than **8 ft** to the edge of a swimming pool. KEEP THEM BACK AND AWAY.

8. Watch your lighting effects so that you do not have neighbor trouble and complaints because your lighting also illuminates his yard. He may want it dark, or he may have some special effect which your stray lighting can spoil!

When you begin planning, it is always appropriate to walk around the area you plan to illuminate, in the daytime. Examine everything and judge what you want to try to bring into highlight and that upon which you just want a low level of illumination. Plan your light locations so that the lighting fixtures and light itself will be as inconspicuous as possible both in the daytime and at night. Examine the areas in the early morning and late evening twilight when the shadows are long and strong. You can get many ideas for illumination from these kinds of effects from the natural lighting changes. When you walk around, notice how you tend to go from place to place; that is, you don't just watch the ground every minute, but you sort of determine the next point you will reach and go toward that. This means that, while natural hazards should be illuminated in the yard, you should not try to illuminate or floodlight the **entire area.** Light up "point" areas only. One fine type of light fixture for hazard illumination consists of a slotted plate which beams the light down. You've seen this in movie houses, in the aisles. It is effective and yet is extremely inconspicuous.

The next effort in your plan should be to select some area focal points and highlight these, and install all other lighting to gradually vanish to darkness or blend into another focal area. You must determine the highlight areas **first** in your projected lighting plan. Remember, you do not want to compete with the landscape and architecture through the use of conflicting colored lights. Use of colored lights is sometimes not recommended unless careful trial confirms that it is an advantage and adds to the effect. Colors are primary and usually harsh. **Red** does not enhance **green** growth; **blue** is eerie, some say, but **can** add depth; and your **green** may or may not be the correct color to enhance shrubbery. If you try color, do it carefully and check the effects.

Color Planning

Let's examine the color phenomena further. As you are aware, your television set uses only three primary colors to attain all shades and hues. The primary colors are red, blue and green. The selection of these colors and the manner in which they blend to produce other colors is shown in the Maxwell color triangle, Fig. 6-13. What this means is that you might experiment with cross lighting, that is, the mixing of the beams of different colored lights to see what effects and colors you might achieve.

Another possibility of the color study is to refer to the artist's paintings, or the stage or theatre or the color movies. Notice how one color may be imposed on a background of

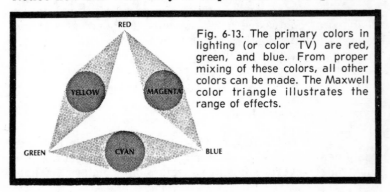

Fig. 6-13. The primary colors in lighting (or color TV) are red, green, and blue. From proper mixing of these colors, all other colors can be made. The Maxwell color triangle illustrates the range of effects.

Table 6-1. Emotion-Producing Colors

COLOR	EMOTION
Red	Danger
Orange	Warmth and excitement
Yellow	Contentment
Pale green	Kindness
Green	Macabre
Blue-green	Sinister
Blue	Quiet depth of feeling
Violet	Delicate emotion
Cerise	Deep affection
Lavender	Wistfulness

another color to achieve a rare effect. A dark color surrounded with a lighter color makes the darker color still more mysterious. A blue background brings out green by the contrasting effect between the two colors. The ladies are much more color conscious than the men, usually, so here is an area where the man of the family can work at the installation physically, and the ladies of the house can help with their artistic touch and esthetic perception. Table 6-1 will guide your efforts.

There are some generally accepted ideas on use of color. Red or pink colors tend to enhance the human complexion. Test out your color TV by turning the hue (tint) control toward the green-blue and notice the change from the warm and healthy complexion when the reds and pinks are present, to a sickly, mottled presentation. Plan then for pink or reddish colors for party or family areas where people want to appear handsome to each other.

When in doubt about color effects, use **white**. This will at least allow the colors which do exist to be seen and not be changed or contrasted too drastically if the wrong lighting color is used. But don't be afraid to experiment. Remember, if you serve guests under outdoor colored lights, the food should be examined **under this lighting** before the guests arrive. Some colors make even good food unappealing.

System Planning—Hardware

Now we want to examine some actual installations and equipment used. As a preliminary to this, we point out that you may make several kinds of reflectors or "pointers" of light which will confine the beams to the areas you have chosen. Tin (aluminum is better because it does not rust) may be used, and you may mount weatherproof light fixtures inside. Check your local hardware store or light counter in the department stores for types of outdoor fixtures you can put together. Sockets, wires, and tape are all available there, as well as conduit boxes, etc. Several fixtures and accessories made by Loran, Inc., using low voltage for the lighting system, are shown in Figs. 6-14, 6-15, and 6-16. (Loran Products, Inc. 1705 East Colton Ave. Redlands, CA 92373.)

We think that the low voltage systems are best because they minimize the danger of shock and hazard due to the voltages of the normal house system. The electricity is isolated by a transformer. This means you won't get shocked if you accidentally touch one side of a low voltage live wire and you are on the ground. There is **no ground return** for this system as there is with the normal electrical input to your home. The transformer reduces the voltage from 120V to 12V and is necessary for continuous use of lights.

So the second part of the project, after determining **what** you want to light and roughly how you will begin the project— that is, what kinds of lighting you will first try—is to locate the exterior centers of wiring distribution and plan for **transformer installations** at those points. Notice that you must run the house current from the house to the centers of distribution.

Fig. 6-14. Some types of low voltage lightscaping lamps and devices available from Loran, Inc. The range is from long, narrow beams (on the upper left) and circular diffused spots, upper right, to the tightly beamed spots in cylinders as shown in the row below. Those on the left are for uplighting and those on the lower right are for downlighting.

Fig. 6-15. Area illumination fixtures, upper row, and the low voltage transformers, lower row, are used in the low voltage type lightscaping of homes and gardens. (Loran, Inc.)

Proper underground cables or conduit are required, Figs. 6-17 through 6-24, and proper fusing is mandatory. If 120V is used outside, remember to have a GROUND FAULT INDICATOR (Chapter 8) INSTALLED BY YOUR ELECTRICIAN.

The transformer should be mounted above ground. It must be weatherproof. Plan to keep all cables to lights of No.14 and 18 wire less than 100 ft long. Plan each transformer as the hub of a wheel with the wiring radiating out to the various lights as spokes. Make the fixture runs of wire as direct as possible. The longer a wire is, the more hindrance to current flow, and possibly a reduction in the light intensity. Mount the transformers at least 1 ft off the ground. Do all your wiring in the daytime and be sure to disconnect all primary power from the house during this phase of the installation. Use type "UF" cable for the underground runs. You merely dig a thin slit trench to put the wire in and dig down so the cable is about 6 to 8 in. below ground level so normal cleaning and surface soil movement will not affect it; see Fig. 6-24.

Now, when evening comes, turn on your lightscaping system and check the effects and locations. Experiment with bulb size and placement and method of lighting. Once you are satisfied, you may make the installation more or less permanent. You will probably want to change some things later,

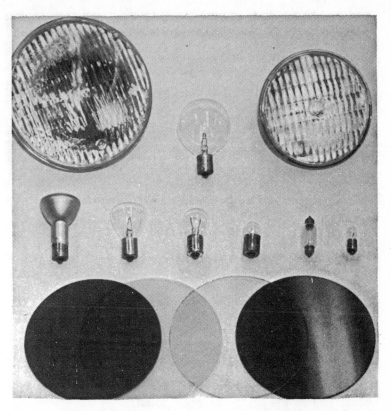

Fig. 6-16. Some types of lamps used in exterior lighting of homes and offices. Note the color discs in the lower row. (Loran, Inc.)

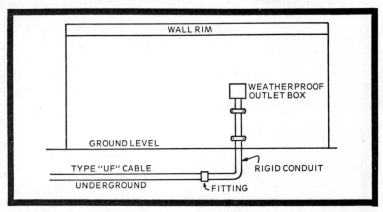

Fig. 6-17. The type of cable, and the method of placing an outlet box on the wall of the house, or fence. Note that the type "UF" cable must be joined through a fitting to rigid conduit.

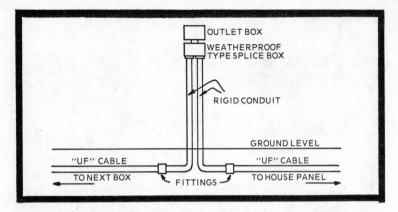

Fig. 6-18. Method of adding outlet boxes around the patio or in areas of the yard. Note again the requirement that the "UF" cable which is run underground must be connected to rigid conduit to support the outlet box. Note also the splice box which is used to make connections so that the lines can continue to other outlets.

perhaps seasonally, just to obtain different views and effects. You can now install your timers, light controlled switches, dimmers, etc.

EXTERIOR WIRING CHECKLIST

Here are some technical specifications for **outdoor** wiring which should always be borne in mind.

a. Is the service entrance (the input to the house from the

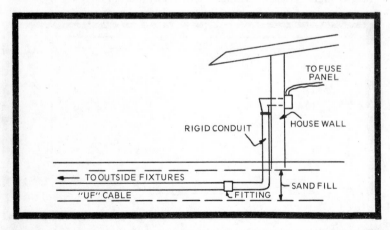

Fig. 6-19. The approved method of burying a cable underground. Note sand fill for drainage.

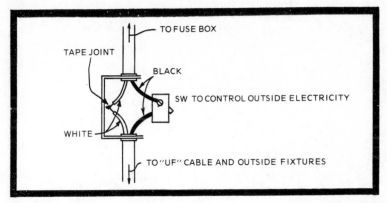

Fig. 6-20. Method of connecting a switch in the rigid conduit cable. The switch box must be a weatherproof type if outside.

electric company line) large enough to handle the additional load of the outside wiring? This can be checked against the total loading of the house and the size of the service wires. Remember the amperage table on wiring? (Refer to Table 1-1.) Always consider the LENGTH of lines to be run, as the voltage will be reduced if the line is long and the wire size is

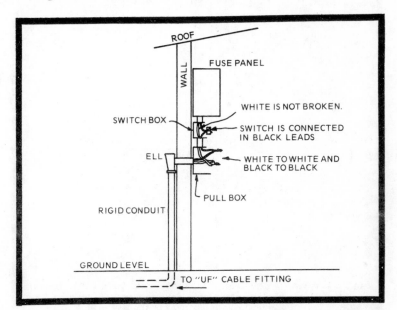

Fig. 6-21. Method of running the cable from the house out to the yard area. A protective switch may be placed in the line using this type of circuit wiring. Note that the switch would be in the **black** leads.

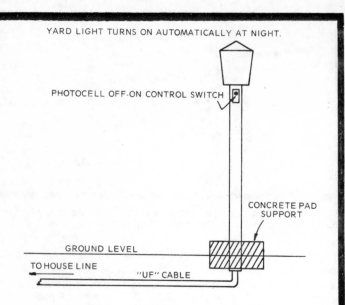

YARD LIGHT TURNS ON AUTOMATICALLY AT NIGHT.

PHOTOCELL OFF-ON CONTROL SWITCH

CONCRETE PAD SUPPORT

GROUND LEVEL

TO HOUSE LINE

"UF" CABLE

Fig. 6-22. The yard light is simple to install.

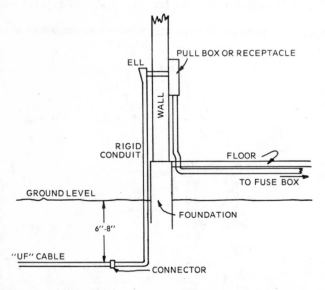

ELL

PULL BOX OR RECEPTACLE

WALL

RIGID CONDUIT

FLOOR

TO FUSE BOX

GROUND LEVEL

FOUNDATION

6"-8"

"UF" CABLE

CONNECTOR

Fig. 6-23. When it is necessary to come through the house through the back of an existing receptacle box, this method can be used. You could simply drill through the back of the box in the house and run the "UF" cable in through this hole. Then connect to the existing receptacle wiring, or use rigid conduit as shown.

too small. Review the chapter on interior wiring for wire size and length of runs.

b. Consider whether spare circuits are available at the fuse panel inside the house from which to run an extension, or if an additional fuse panel is needed.

c. Consider having convenience outlets of 120V installed around the patio area, with **GROUND FAULT INDICATORS** in each 20A circuit.

d. Any circuit installed in an existing dwelling for outdoor use should be a new circuit originating at the panel (fuse) board rather than an extension of an existing circuit. This is because the loading for the interior circuit was planned when the house was built. If it is an older home, probably this circuit is somewhat near the peak of its capacity now or perhaps it is even overloaded. Adding still more demand to it is DANGEROUS! In all cases then, the new outside wiring circuit originating at the panel (fuse box) should be for this purpose only. You may want a qualified electrician to install these primary circuits.

e. When you plan the outside circuits (except for those using low voltage) consider this: Once there is available an outside socket or terminal or receptacle with the **normal house voltage**, you and the family will always find things to plug into it! Tools and appliances will just seem to be made for use in such places, even if they had not been considered there before!

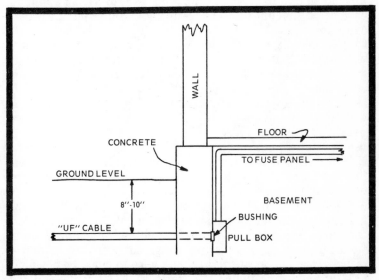

Fig. 6-24. When it is desired to go through the foundation, this method is used.

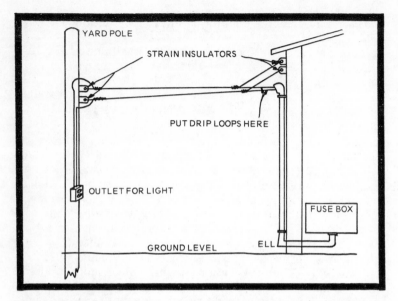

YARD POLE

STRAIN INSULATORS

PUT DRIP LOOPS HERE

OUTLET FOR LIGHT

FUSE BOX

GROUND LEVEL   ELL

Fig. 6-25. Sometimes it is necessary to use overhead wiring to get electricity into the yard area. This is particularly true in farm situations. Be sure to add drip loops where the wires from **out** of the house conduit line.

So be sure your circuits are large enough to provide for this. Have 20A circuits, and at least one or two 120/240V **3-wire circuits** so that if, at some time, you want to use an appliance or tool or machine which needs 240V, this voltage will be available.

f. Most installations of outside wiring will be either from a basement with its concrete walls, or cement block, or from the ground level where the panel is located in the garage or some other point on the first floor of the house. The wiring can be brought to the outside as was shown in Figs. 6-21, 6-23, and 6-24.

g. If it turns out that it is advantageous to run the wiring **overhead** from the house to a central distribution point, then use the same method of getting outside; that is, use conduit to run the lines to a high enough level **on the side of the house**, and then use proper insulator strain fasteners to hold the wires overhead as in Fig. 6-25.

h. Remember that for underground runs, the type "UF" cable is approved for direct burial in the earth. Dig a trench deep enough so that normal spading will not hit the cable. Be sure stones and sharp objects are removed from the bottom of trench. Use a layer of sand around wire to enhance drainage.

i. It is necessary under the National Electrical Code that underground cables be continuous lengths from terminal to terminal . Plan for this. DO NOT TRY TO TAP AN UNDER-GROUND CABLE. IT'S DANGEROUS AND ILLEGAL.

j. If you have animals or insects which might attack the cable, then use armored cable or conduit, or lead-sheathed cable. Confer with your electrical dealer or department store electrical counterman.

k. On all overhead wiring be sure to use "drip loops" where wires go out of or into buildings, or conduit. For runs over walls, roofs, etc., the clearance is recommended at 8 ft; over driveways, etc., 12 to 18 ft; over walks at least 10 ft. And keep the wiring at least 3 ft away from windows and doors.

l. For runs of up to 20 ft, use No. 14 wire. For runs from 50 to 200 ft, use No. 12 wire. For runs over 200 ft, use No. 10 wire.

m. Always support the wires so that limbs, wind, etc. will not cause damage to them.

Chapter 7

Office Lighting

The office is where the money is made! The better the lighting, the better the impression given the customer. Also, the better the lighting, the fewer mistakes, and the more efficient the office staff. That's money in the bank also! The name of the game is to improve efficiency, reduce errors, increase employee morale, present a good public image, make the customer happy, and **make more money!**

Let's think over an example. If you have bad lighting which causes 10 minutes a day lost time or if you employ 50 people who are so affected, this means 500 man-minutes per day lost time. This is 8.3 hours. Each day, you are paying for a phantom employee who does nothing! Good lighting eliminates such ghosts. And, of course, no one need tell you the value of a good public image. Examine Fig. 7-1 for an example of an impressive atmosphere.

People respond to their surroundings physically and emotionally. In good office design, both aspects must be planned for, since people like to work in areas that are well lighted and have interesting appointments in pleasing colors. In such environments, employees are able to use their visual ability and manual skills to advantage. And they have a stronger motivation to do so. This, of course, results in more money in the bank. Figure 7-2 shows a pleasing atmosphere.

The principal design factors in such a working environment are visual, sonic, thermal, spatial, and esthetic. Lighting, a key element of the visual environment, is frequently designed into a ceiling system which integrates the visual with the thermal, sonic, and spatial design factors. When this is done, the design is also usually esthetically pleasing. Figure 7-3 illustrates such a case in point.

Research and experience have provided helpful criteria to guide the design of the visual environment. Compromise with these criteria should not be made in areas where **critical** visual work is performed. However, within the indicated design limits there is considerable latitude for the creation of

Fig. 7-1. A total of 112 eight-foot-long cool-white, high output fluorescent lamps provide 100 foot-candles of maintained illumination. The lamps, housed in flush mounted fixtures, feature an acrylic lens. They are spaced on 12 ft centers and mounted 14 ft above the floor. (GTE Sylvania.)

Fig. 7-2. Fifty-seven 400W Metalarc lamps provide 75 foot-candles of maintained illumination. The lamps are housed in Holophane fixtures, feature an acrylic lens, are spaced on 10 by 12 and 10 by 10 ft centers. They are 22 ft above the floor. (GTE Sylvania.)

Fig. 7-3. Recessed ceiling illumination provides bright interior which reduces eyestrain on demanding work. (G.E.)

Fig. 7-4. A pleasant, well lighted office which is informal but efficient. (G.E.)

conformity and productivity. Figure 7-4 illustrates a modern, pleasant, informal type of office. Notice the ceiling panels which provide light without glare and which would not compete with the window if the blinds and drapes were open. Here, the lighting seems to enhance the furniture groupings and contributes to a neat, clean, pleasant atmosphere, which, in turn, enhances efficiency.

DESIGN FACTORS

There is a recommended value of office illumination levels which is important. Examine it in Table 7-1. Recommended foot-candle values are based on:
1. Need (the indications of research).
2. Practicability (the available lamps, equipment, and techniques).
3. Economics (the cost of light balanced with human benefits). Today, economics indicate a compromise below the illumination level needed for best performance and below that

Table 7-1. Recommended Illumination Levels for Various Office Tasks and Locations.

Area	Minimum foot-candles on the task
General Offices	
General office work	10
Accounting, bookkeeping	150
Drafting, designing, cartography	200
Private offices	100
Conference rooms	½ to 100 *
Corridors, elevators, escalators, stairways	20 **
Lobbies, reception areas	30 ***
Washrooms	30

  * More than one lighting system or dimming control desirable.
 ** Preferably 1⁄3 to 1⁄2 the level of nearby offices.
*** May be more when planned for transition with daylight or other areas.

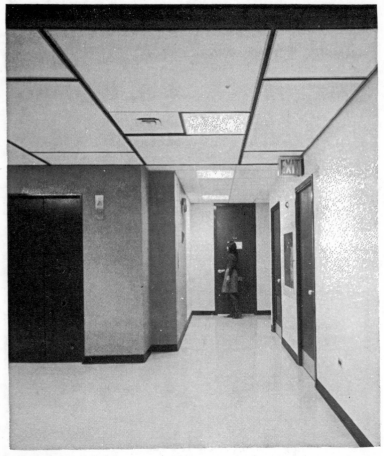

Fig. 7-5. An ultra modern "square ceiling" effect obtained with Sylvania Curvalume lamps. The lamps are installed in 2 x 2 sunbeam fixtures in the ceiling and provide 150 foot-candles of lighting in office and tabulating areas. (GTE Sylvania.)

practicable to supply with available equipment and application techniques.

A lighting system must also be designed to prevent the illumination from falling below the minimum value. To do this, more light than the minimum is required when the system is new to compensate for dirt collection and lamp depreciation between cleaning and relamping operation.

If the visual task is near a sidewall, luminaires should be spaced closer in that area, or lamps of greater light output should be used. (Uniformly spaced lighting equipment supplies less light near walls.) If a task is on a slanting or vertical surface, less light is available than on the horizontal and additional light must be supplied. Or, when feasible, luminaires with nonsymmetrical light distribution can be used.

One example of the hallway lighting suggested is shown in Fig. 7-5. Note that it is inconspicuous, yet the level is sufficient to give a bright, comfortable atmosphere.

Now let's talk for a moment about the benefits of the well designed office environment, considering these from a higher supervisory level. The **competent executive** knows that a well designed office results in more job applications from a better class of worker. This means substantially upgrading the caliber of personnel of the office in fewer years. The best executives know that longer average tenure of employment results from having attractive office spaces. Reduced turnover can save thousands of dollars annually because of less need to train new personnel.

Less fatigue means increased **quality** of work, and **quantity** of work is a natural follow-on to this. The employee who does not suffer eye fatigue works better all day long. Of course, there are fewer mistakes in this case also. That great intangible, but such an important factor, is **employee morale**. Good office environments affect the employee's emotions positively—his feelings toward his work, his boss, and his company. Of course, this applies to the women workers as well.

Finally, there is the greater prestige for the organization because of the impression that a well managed, well decorated, well lighted office gives to others. It has been estimated that a typical office, with everything in it and considering wages paid, costs about $64 per sq ft per year. The lighting that supplies the minimum recommended levels of illumination costs only about 1 percent of this figure. This is a small price to pay for the value received.

As a further example of the effect of good lighting in some commercial environments, see Figs. 7-6 and 7-7 for the proper manner in which to illuminate conference rooms. Of course, consideration must be given here to provisions for audio-visual equipment displays, and controlled levels of lighting for this purpose.

In Fig. 7-8 is shown the lobby illumination for a large modern office building.

TECHNICAL CONSIDERATIONS

Consider now some technical aspects of office lighting. The basic premise is **efficiency**, and proper levels of lighting

Fig. 7-6. A well designed conference room, with proper levels of illumination (G.E.)

Fig. 7-7. A large conference room needs overall **and** area lighting. Notice the brightness of the desk areas. Overlapping light is provided through the 30 in. modules. (GTE Sylvania.)

as we look at it from this viewpoint. Let's examine the requirements for a private office.

The functions of private offices can include the most demanding of visual paper work, casual conversation with visitors and associates, social activities, and entertaining. Lighting should be planned with the flexibility needed to set the mood for any function. The lighting system must provide, however, illumination suitable for the most difficult visual task commonly encountered.

Small rooms utilize light less efficiently than large rooms. Because of this, spacing between luminaires must be closer than in general offices to provide the same illumination level. This can also be accomplished by using higher light output lamps in smaller offices.

Where the desk is in a fixed location, lighting equipment can be positioned where it will produce more illumination on the desk top than elsewhere in the room. Vertical surfaces

Fig. 7-8. A beautiful office building interior. Some 20,000 lamps used in the office areas are housed in 31 in. square fixtures. The recessed lighting troffers have complete air-handling capabilities. This is the executive area of the new 15 million dollar headquarters of the Boise Cascade Corporation in Boise, Idaho. (GTE Sylvania.)

should not be neglected, however, since they can become dark and visually uncomfortable with nonsymmetrical lighting. Enivironmental lighting can create a more pleasing visual surrounding as well as a more confortable one. Lighting for the desk should generally be centered over the occupant's head rather than over the desk so that light source reflections in the seeing task are minimized. Incandescent downlights should not be used over a working area because they produce harsh shadows and bright reflections.

The visual environment for **general offices** is of extreme importance because of management's investment in the performance of people who work there. Even though lighting in an existing environment may seem adequate, new lighting that provides significantly more illumination of good quality will often stimulate improvements in employee performance worth far more than the cost of the system.

Fluorescent lamps are the most practical source of general illumination for offices. Among the lamps available, the GE Power Groove fluorescent lamp offers lower initial lighting cost because there is more light per lamp, and fewer lamps and fixtures are needed.

The choice of fluorescent lamp **color** involves (1) efficiency, (2) color-rendering properties, and (3) atmosphere or type of "whiteness" created by the lamps. The most appropriate colors of fluorescent lamps for offices are listed below.

1. Cool White: Most widely used; high efficiency; daylight type atmosphere; use for most working offices.
2. Deluxe Cool White: Excellent color (like natural daylight); 30 percent lower efficiency; for best appearance of people and furnishings.
3. Warm White: High efficiency; warm atmosphere (may be too warm at high lighting levels).
4. Deluxe Warm White: Excellent color (like incandescent lamps); 30 percent lower efficiency; for reception areas, lounges, cafeterias, etc.

Tinted lamp colors should not generally be used as they distort some colors and provide no particular seeing benefit.

Incandescent lamps are only about one-third as efficient as fluorescent, and indirect lighting is the only type of incandescent illumination suitable for working areas. However, due to lower efficiency, incandescent lamps cannot economically provide the lighting levels required for good office performance. **Downlighting** should never be used over desks because of disturbing shadows and reflections.

More general rules for good lighting design in offices can be listed as follows:

a. Use appropriate reflectances on walls and ceilings. The room and work surfaces should have reflectances within the ranges of Table 7-2, to obtain the brightness relationships essential for visual comfort. Literally hundreds of colors are available within these ranges to provide for desirable psychological and esthetic reactions. Reflectances outside the recommended ranges may provide dramatic effects, but they may open the door to visual discomfort and complaints of various kinds later. Accent colors are desirable in **small amounts**, but never should be present in the area surrounding the immediate task. Wall lighting may provide illumination which is suitable with various degrees of reflectance values. A method is shown in Fig. 7-9.

b. The general overhead lighting equipment is used to light the walls if the wall itself is not illuminated. If the first row of ceiling luminaires is too far from a wall, the wall will be dark regardless of the reflectance color. Spacing between a row of luminaires and a parallel wall should not exceed half the spacing between rows of luminaires. The luminaire considered here is a fluorescent fixture flush with the ceiling. The ends of the rows of fluorescent luminaires should come to within 6 to 12 in. of the wall. If desks are located next to a wall, the distance between the wall and the parallel row of luminaires should be no more than 30 in. maximum. When curtained windows are illuminated, the method of Fig. 7-10 should be used.

Table 7-2. Recommended Reflectances for Surfaces in Offices

	Reflectance	
	INDEX	PERCENT
Ceiling finishes	0.80	15 *
Walls	0.50	20
Furniture (desk tops, table tops, etc.)	0.35	25 **
Office machines and Equipment	0.35	25 **
Floors	0.30	30

* For finish only. May be lower.

** American standard.

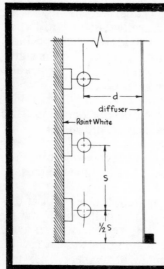

Fig. 7-9. The method of installing a "luminous wall." For uniform brightness with high diffusion transmitting materials, S : d ratios equal to or less than 1.5 are used. The spacing may be reduced between lamps and diffuser to increase brightness also.

c. Always keep the spacing between ceiling luminaires small. The luminaires should be so installed that they permit considerable overlap of the light distribution. This will minimize troublesome shadows and ceiling reflections. Lighting quality is improved if the spacing between luminaires is within the maximum spacing-to-mounting height which is provided with each kind of fixture for a multiple fixture installation. The higher the proportion of **direct** light from the fixture, the closer the spacing should be to gain overlap.

d. Always consider the proper **Visual Comfort Probability** (VCP), a comprehensive method of assessing visual comfort. It takes into account luminaire brightness, room size, ceiling height, illumination level, and room reflectances. It indicates the percentage of people who would appraise a lighting installation as comfortable from a center, rear-of-room location. Although only one measure of the quality of a lighting system, VCP is important in preventing eye-jarring luminaire brightness. The VCP has been approved as an appropriate method of evaluating direct glare by the Illuminating Engineering Society (IES), and many fixture manufacturers now have the VCP figures available for their equipment.

e. You should always be sure that undesired reflections are minimized. Specularity (reflection) can cause fatigue and errors, etc.

Specularity exists to some degree in many common office materials, including pencil writing, printing and writing inks, paper, etc. If a light source or window is reflected at the angle

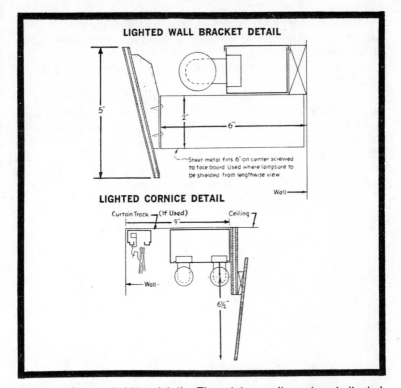

LIGHTED WALL BRACKET DETAIL

5"

2"

6"

Sheet metal fins 6" on center screwed
to face board Used where lamps are to
be shielded from lengthwise view

Wall

LIGHTED CORNICE DETAIL

Curtain Track (If Used)   Ceiling

9"

Wall

6½"

Fig. 7-10. Cornice lighting details. The minimum dimensions indicated
are important for good lighting results with these two techniques. For a
smooth brightness gradation from ceiling to floor, the fluorescent light
source should be 1 ft away from the wall for each 4 ft of vertical surface to
be lighted. Spacing between incandescent lamps should be not more than
their distance from the wall. Incandescent reflector and projector lamps
are available ranging from 30 to 500W to suit various spacing relationships. Approximately 75W per ft in incandescent lamps are desirable for
lighting a 35 percent reflectance wall in a room with 100 foot-candles of
general illumination. About 25W per ft are desirable when fluorescent
lamps are used in reflectors.

of viewing, the contrast of the task may be reduced. To
minimize such veiling reflections:
1. Eliminate specular (glossy) surfaces in the task and
surroundings. Specify matte finishes for furniture, machines,
stationery, business forms, etc.
2. Use low gloss inks and ballpoint pens in preference to pencil
writing.
3. Light the task from several directions by close spacing of
luminaires.
4. Orient viewing parallel to, and work positions in between
rows of luminaires.

If the above suggestions are impractical, the general lighting should be provided by luminous indirect luminaires or a luminous ceiling.

f. You must control window glare by controlling the lighting level of the daylight situation.

To maintain adequate visual comfort, the same brightness limits should be applied to window walls as other walls. Regardless of window orientation, shielding of daytime brightness or nighttime darkness is necessary. The visual design should incorporate draperies, blinds, shades, glazing with reduced light transmission, building overhangs, or a suitable combination of such techniques.

Practical Wiring Tips & Safety

This chapter gives you some tips and diagrams which may be helpful to you when making electrical additions or original installations. When making additions to the wiring, you must make certain that the service is large enough to support the additional current demands. Examine the service input, Fig. 8-1.

WIRE & CONDUIT SIZES

The size (gage) of the entry wires will govern the amount of current which can be drawn inside the house or building. Notice the method of getting the wiring through the meter and into the fuse panel. Notice also the common ground wire which is "fed" through the back of the meter case. The size of the input conduit varies with the load expected. A ¾ in. conduit is used for three No. 8 wires, a 1¼ in. for wires from No. 2 to 6 when three of this size are used; No. 1 wires requires 1½ in. conduit, and a 2 in. conduit is used for three No. 1/0, or three No. 2/0 or three No. 3/0 wires. At the point where the conduit goes through the wall you will find an entrance ell, as it is called. This is a right angle fitting with a removable back so that wires can be pulled down and then out and reinserted to go through the wall.

To determine the service ampere capacity, you must know the wire size. It may be marked on the insulation, or you may have to obtain a wire gage which will indicate, by the physical size of the conductor, the number of the wire. You may be able to ask for, and obtain for a few cents, samples of each of the above sizes of wire from the store and make a direct comparison with your service wire. Be sure to compare the **metal conductor** part, and not the overall size with the insulation on it. Insulation thicknesses will vary. Also, even if the sample you measure is a solid single conductor, and your service wires in your panel at home are stranded conductors, the relative size will be practically the same—the stranded

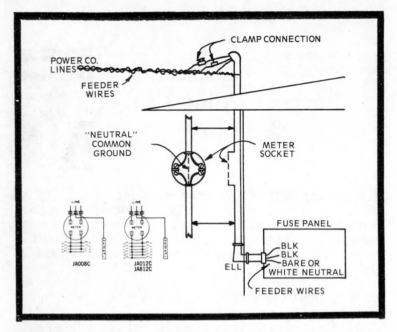

Fig. 8-1. Details of the power service input to a home. Two meters may be used, one for the water heater alone.

may be just a slight bit larger physically, for the same capacity. But you will be able to tell, quite easily, the difference between sizes from No. 2 to 1/0, for example.

FITTINGS

When the wires leave the fuse box, they run through conduit, or are in nonmetallic sheathed cable. The conduit or cable terminates in outlet boxes, or fixture boxes, such as are shown in Fig. 8-2. There are various types of fittings used to terminate the conduit or cable in these boxes (Fig. 8-3).

When connecting wires inside a box, either to a fixture, switch, or to extension lines which go on further from the box, use the connecting method shown in Fig. 8-4. Notice that all black wires connect together, and all white wires connect together. If the wiring runs to a switch and then to a light fixture, the method is as shown in Fig. 8-5. Here, a **red** and white wire go up to connect to the light fixture wires. A method of supporting a fixture on a fixture stud is shown in Fig. 8-6.

Two methods of connecting the switches are shown in Figs. 8-7 and 8-8. Figure 8-7 has a rigid conduit, and Fig. 8-8 shows how a nonmetallic cable is used.

TESTING

A method of testing a circuit such as shown in Fig. 8-9 is to find out which lead is the "hot" lead to a switch—just in case someone wired a switch without using the proper color code. Open the switch box as shown, use a lamp—with its plug removed—as a test instrument; connect one lead to the switch connection and the other lamp lead to the metal box. When the power from fuse box to this circuit is on, the lamp will light if you touch the "floating" lead to the hot side of the switch. The other side of the switch wire will then be known to go to the light.

ADDING OUTLETS

Sometimes you desire to add a new outlet by going through the back of a switch, through a wall, to a new box and outlet receptacle. You can do this as shown in Fig. 8-10. You can then install a lamp in the new area and use a plug to connect it to the house current. You will have to use the method of identifying the "hot" switch leg as was shown in Fig. 8-9—then the two wires would be the white wire tap, and a line from this "hot" side of the switch. BE SURE THE MAIN CIRCUIT BREAKER IS OFF WHEN YOU MAKE THESE CONNECTIONS!

Sometimes you want to get an additional outlet near a switch. You can change the switch fixture itself to a new set

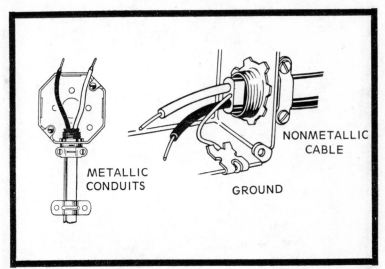

Fig. 8-2. Method of connecting rigid conduit and nonmetallic sheathed cable into outlet or terminal boxes.

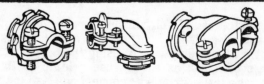

Fig. 8-3. Some types of cable termination fittings. There are many types—ask your dealer about them.

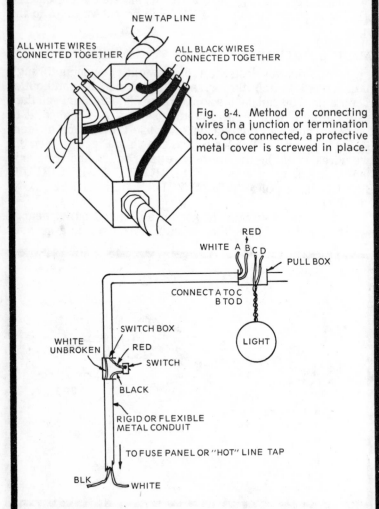

NEW TAP LINE

ALL WHITE WIRES CONNECTED TOGETHER

ALL BLACK WIRES CONNECTED TOGETHER

Fig. 8-4. Method of connecting wires in a junction or termination box. Once connected, a protective metal cover is screwed in place.

RED

WHITE A B C D

PULL BOX

CONNECT A TO C B TO D

LIGHT

SWITCH BOX

WHITE UNBROKEN

RED

SWITCH

BLACK

RIGID OR FLEXIBLE METAL CONDUIT

TO FUSE PANEL OR "HOT" LINE TAP

BLK

WHITE

Fig. 8-5. Some details of connecting a switch and a light fixture. Notice that the **white** wire must **not** be broken by the switch.

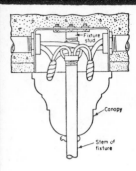

Fig. 8-6. One method of supporting a fixture with a fixture stud which screws into the base of the wire termination box. Note how wires from the fixture are connected to the "line" wires by taps.

Fig. 8-7. Installation of a toggle switch in a box with rigid conduit connections. Note plastic insulation cap on connection. (Arrow-Hart.)

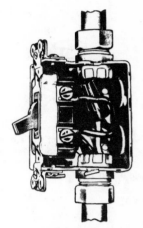

**Cutaway Box
Showing Wire Room**

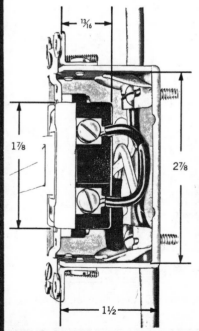

Fig. 8-8. Installation of a switch in a box which is connected to flexible nonmetallic sheathed cable. Note the clamps for the cable. When you use this type of wiring you should provide for a **ground** wire to ground all metal boxes and if the cable does not have this 3rd wire, then you should run a separate **green** grounding wire along with the cable. (Arrow-Hart.)

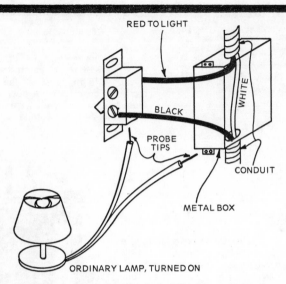

Fig. 8-9. Method of determing the hot side of a switch when wires are discolored, are the same color, or when there is doubt for any reason whatever. Normally the hot wire is black, the cold or switch leg is **red**.

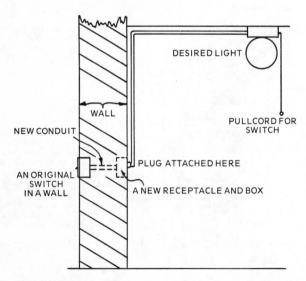

Fig. 8-10. Method of adding a new outlet in the wall opposite a switch or existing outlet.

consisting of an outlet, a switch, and a light (or another small outlet) if you desire. The kind of equipment needed is shown in Fig. 8-11. The switch would be connected exactly as the original switch was, and extension wires from the "hot" side of the switch, and a tap into the neutral wire will furnish the two wires needed to connect to the receptacle terminals so they will always be ready to provide power regardless of whether you have the switch on or off. Just in case you connect this equipment and find that when you turn the switch off the receptacle power goes off, you will know that you have made a connection from the WRONG side of the switch itself to the receptacle.

Locating Studs in Walls

Once in a while you want to locate a stud in the wall so that you can then cut in close to it to install a wall box, such as in Fig. 8-12. Of course, there are many, many types of wall box mountings, some with metal hangers for mounting in between studs, and others which have expansion sides for mounting directly into a wallboard finished wall. In any event, there is a method by which you can locate these studs without just making a million holes in a row, or tapping with your fist to determine the hollow and solid sounds. In Fig. 8-13 you see how a hole is drilled into the wall and a wire is run into the hole; since it is bent, it can be turned so that you can feel when it strikes a solid joist (stud). Measure the length of the wire by holding it tightly as you withdraw it, and you'll have the distance to the joist. You can then find the joist in the opposite direction and from the distance between the two, locate all other joists in the wall, since they will have the same spacing.

Also, it has been said by some that you can use a dime-store compass to locate joists. Since nails are used along the baseboard into the joist, run the compass slowly along the baseboard. The needle will deflect in the presence of the nail, and that locates the joist.

Safety Bonding Outlets

Next we examine, in Figs. 8-14 through 8-17, some methods of connecting outlet receptacles of the **safety bonding type**, made by Arrow-Hart Co., to various types of conduit and cable.

Inside Surface Wiring

General Electric makes a fine type of cable suitable for inside surface wiring. This is a good type to use to replace

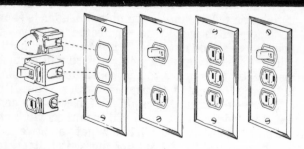

Fig. 8-11. Some small receptacles and switches which can be used to replace the larger switch. This gives an added outlet receptacle easily.

Fig. 8-12. Some common types of outlet boxes. The juncture or termination or "pull" box is usually 6-sided or round.

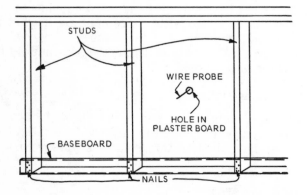

Fig. 8-13. A view of the nails which fasten the baseboard to the studs

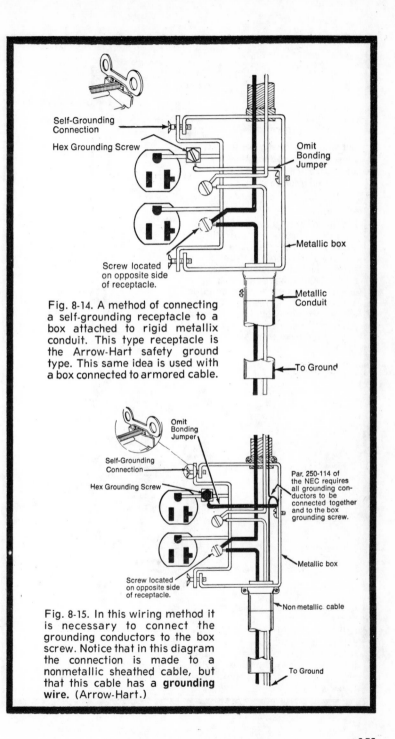

Self-Grounding
Connection

Hex Grounding Screw

Omit
Bonding
Jumper

Screw located
on opposite side
of receptacle.

Metallic box

Metallic
Conduit

To Ground

Fig. 8-14. A method of connecting a self-grounding receptacle to a box attached to rigid metallix conduit. This type receptacle is the Arrow-Hart safety ground type. This same idea is used with a box connected to armored cable.

Omit
Bonding
Jumper

Self-Grounding
Connection

Hex Grounding Screw

Par. 250-114 of the NEC requires all grounding conductors to be connected together and to the box grounding screw.

Metallic box

Screw located
on opposite side
of receptacle.

Non metallic cable

To Ground

Fig. 8-15. In this wiring method it is necessary to connect the grounding conductors to the box screw. Notice that in this diagram the connection is made to a nonmetallic sheathed cable, but that this cable has a **grounding wire.** (Arrow-Hart.)

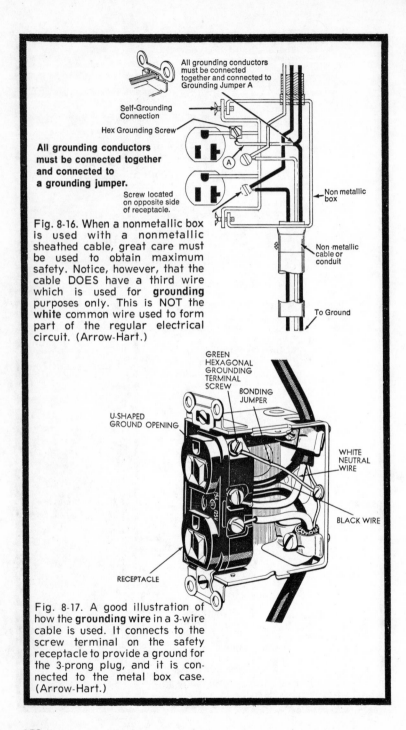

All grounding conductors must be connected together and connected to Grounding Jumper A

Self-Grounding Connection

Hex Grounding Screw

All grounding conductors must be connected together and connected to a grounding jumper.

Screw located on opposite side of receptacle.

A

Non metallic box

Non metallic cable or conduit

To Ground

Fig. 8-16. When a nonmetallic box is used with a nonmetallic sheathed cable, great care must be used to obtain maximum safety. Notice, however, that the cable DOES have a third wire which is used for **grounding** purposes only. This is NOT the **white** common wire used to form part of the regular electrical circuit. (Arrow-Hart.)

GREEN HEXAGONAL GROUNDING TERMINAL SCREW

BONDING JUMPER

U-SHAPED GROUND OPENING

WHITE NEUTRAL WIRE

BLACK WIRE

RECEPTACLE

Fig. 8-17. A good illustration of how the **grounding wire** in a 3-wire cable is used. It connects to the screw terminal on the safety receptacle to provide a ground for the 3-prong plug, and it is connected to the metal box case. (Arrow-Hart.)

"extension cord" type additions to your home wiring. Notice in Fig. 8-18, how the cable may be installed using the nails provided, and how the cable itself is so designed that the nail holes are already incorporated. The kinds of surface fixtures which can be used easily with this kind of cable are shown in Figs. 8-19 through 8-21.

Methods of connecting various switches, outlets, and lamp holders are shown in Fig. 8-22.

Surface Raceways

In many homes it is desired to have a more permanent type installation such as is made possible by surface raceways. One type is made by the Wiremold Co.; the method of making the installation is shown in Fig. 8-23. Many types of

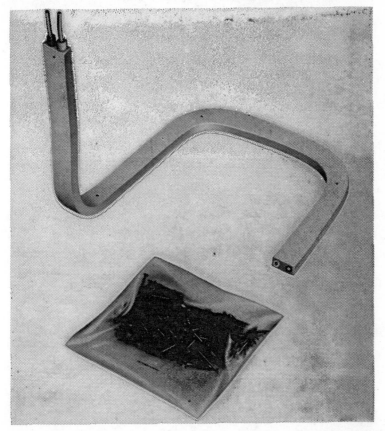

Fig. 8-18. General Electric's flexible interior surface wire. This is easy to install and durable.

Fig. 8-19. A surface mounting receptacle.

Fig. 8-20. A surface mounting type of lamp fixture.

raceways are available, and some examples are shown in Fig. 8-24. Note that the base is fastened to the surface and the wires are laid in place and secured with small clamps; then a top is snapped into place, making a very neat, durable, and safe installation. The manner in which these raceways may be mounted is shown in Fig. 8-25. Some ideas here are suitable for mounting other types of equipment such as outlet boxes.

With the raceways are various fittings and attachments; outlets may fit directly into the metal cover, as also will switches and lamp receptacles. Corner fittings are shown in Fig. 8-26, and channel installations in Fig. 8-27.

Professional electricians, of course, know exactly how to make the neatest and most practical installations (Figs. 8-28 through 8-30). In Fig. 8-30 notice how the installation blends into the room decor. The baseboard installation is hardly noticeable. This is a fine way to add new circuits and outlets in your home without detracting from the room beauty, and without the expense of going inside the walls, ceilings, etc.

SWITCHING & DIAGRAMS

Next we examine a 3-way switch circuit to turn lights on and off from two locations. The wiring diagram is simple (Fig. 8-31), and the switch itself is shown in Fig. 8-32.

More switch wiring diagrams are included for your information in Fig. 8-33, while Fig. 8-34 shows the multiple receptacle outlet drawings and some wiring diagrams for the units which can be used in these kinds of installations. Remember that you can replace a regular size switch or receptacle with these small multiunit type fixtures without changing the box size. Many times it's an easy way to get more outlets or switches in a room with the least amount of additional wiring.

Fig. 8-21. A surface mounting type of switch. This can be used with the units of Fig. 8-19 and 8-20.

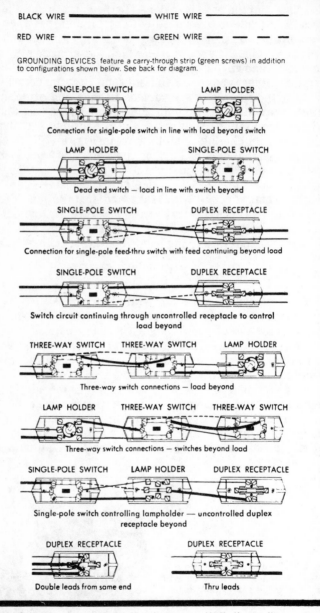

BLACK WIRE ▬▬▬▬▬▬▬ WHITE WIRE ▬▬▬▬▬▬▬

RED WIRE ▬ ▬ ▬ ▬ ▬ ▬ GREEN WIRE ▬ ▬ ▬ ▬ ▬

GROUNDING DEVICES feature a carry-through strip (green screws) in addition to configurations shown below. See back for diagram.

SINGLE-POLE SWITCH   LAMP HOLDER

Connection for single-pole switch in line with load beyond switch

LAMP HOLDER   SINGLE-POLE SWITCH

Dead end switch — load in line with switch beyond

SINGLE-POLE SWITCH   DUPLEX RECEPTACLE

Connection for single-pole feed-thru switch with feed continuing beyond load

SINGLE-POLE SWITCH   DUPLEX RECEPTACLE

Switch circuit continuing through uncontrolled receptacle to control load beyond

THREE-WAY SWITCH   THREE-WAY SWITCH   LAMP HOLDER

Three-way switch connections — load beyond

LAMP HOLDER   THREE-WAY SWITCH   THREE-WAY SWITCH

Three-way switch connections — switches beyond load

SINGLE-POLE SWITCH   LAMP HOLDER   DUPLEX RECEPTACLE

Single-pole switch controlling lampholder — uncontrolled duplex receptacle beyond

DUPLEX RECEPTACLE   DUPLEX RECEPTACLE

Double leads from same end   Thru leads

Fig. 8-22. Some surface line wiring diagrams. With these diagrams and some of the units shown in previous illustrations, you can add wiring to replace the "permanent type" extension cords which are so dangerous to have in the house.

156

Fig. 8-23. For permanent type additions in the home, the Wiremold surface raceways installed by competent personnel make attractive additions and provide all the outlets desired.

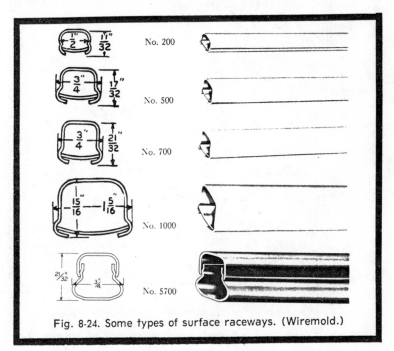

Fig. 8-24. Some types of surface raceways. (Wiremold.)

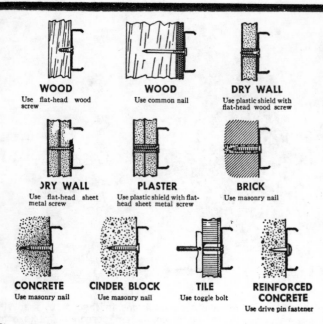

WOOD
Use flat-head wood screw

WOOD
Use common nail

DRY WALL
Use plastic shield with flat-head wood screw

DRY WALL
Use flat-head sheet metal screw

PLASTER
Use plastic shield with flat-head sheet metal screw

BRICK
Use masonry nail

CONCRETE
Use masonry nail

CINDER BLOCK
Use masonry nail

TILE
Use toggle bolt

REINFORCED CONCRETE
Use drive pin fastener

Fig. 8-25. Some methods of attaching the raceways to various kinds of walls and structures.

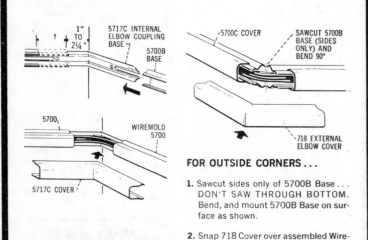

5717C INTERNAL ELBOW COUPLING BASE

1" TO 2¼"

5700B BASE

5700C COVER

SAWCUT 5700B BASE (SIDES ONLY) AND BEND 90°

5700

WIREMOLD 5700

5717C COVER

718 EXTERNAL ELBOW COVER

FOR OUTSIDE CORNERS...

1. Sawcut sides only of 5700B Base... DON'T SAW THROUGH BOTTOM. Bend, and mount 5700B Base on surface as shown.

2. Snap 718 Cover over assembled Wiremold 5700 Raceway sections.

Fig. 8-26. Details of the corner fittings for the Wiremold raceways.

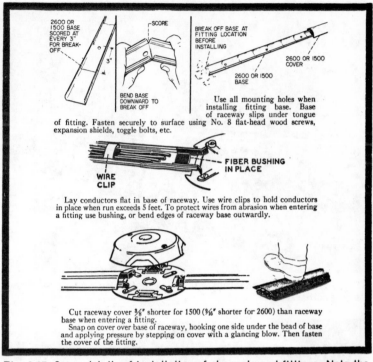

of fitting. Fasten securely to surface using No. 8 flat-head wood screws, expansion shields, toggle bolts, etc.

Lay conductors flat in base of raceway. Use wire clips to hold conductors in place when run exceeds 5 feet. To protect wires from abrasion when entering a fitting use bushing, or bend edges of raceway base outwardly.

Cut raceway cover ⅜″ shorter for 1500 (3/16″ shorter for 2600) than raceway base when entering a fitting.
Snap on cover over base of raceway, hooking one side under the bead of base and applying pressure by stepping on cover with a glancing blow. Then fasten the cover of the fitting.

Fig. 8-27. Some details of installation of channels and fittings. Note the pancake fittings for floor installations.

Fig. 8-28. With care, an installation becomes an attractive new baseboard. Note how the conductors are fitted inside the channel. (Wiremold.)

Fig. 8-29. A multichannel installation is made by placing one raceway atop another. (Wiremold.)

Fig. 8-30. Too add wiring for air conditioners, humidifiers, etc., and at the same time provide for additional outlets, the raceway addition is practical and blends with room decor. (Wiremold.)

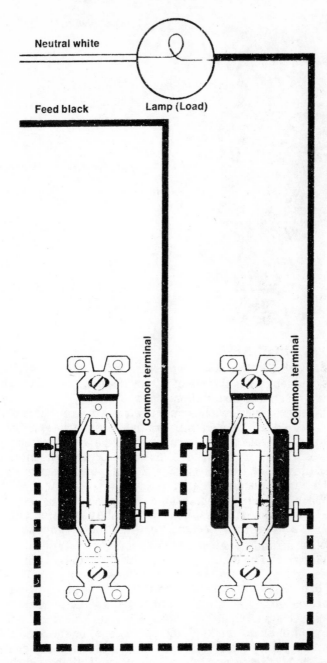

Neutral white

Feed black

Lamp (Load)

Common terminal

Common terminal

Fig. 8-31. The all-important 3-way switch wiring diagram. Either switch will turn the light off or on. The switches may be in two entirely different locations.

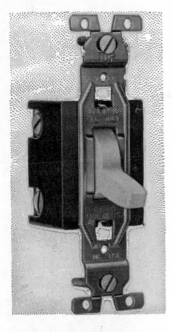

Fig. 8-32. The General Electric 3-way mercury switch is silent and has long life.

SPECIAL PLANNING

The chart in Fig. 8-35 gives some information on voltage drops and lengths and sizes of wires. It's handy to have when making electrical installation plans.

If, when wiring in a basement, wires are exposed to water pipes or placed near damp ground, they are exposed to moisture. It is customary to run wiring **above** any water pipes or pipes which carry moisture. This is to prevent the moisture which may collect on the outside of such pipes from dripping on the wiring conduit or wiring.

When considering bathroom fixtures, remember that metal switch-cover plates should always be used. Metal chains on pull-switches may become "hot" and kill someone.

Now, an additional note or two about the metal wall boxes. Wall boxes can be fastened to tile with through-bolts and fish-plates or toggle-bolts. For concrete, use through-bolts with lead expanders, or expansion bolts. On the wall studs use one of the various type hangers which are available. These are one-ear, center-mounting, etc. Ask your electrical parts counterman about the kind you should use for the kind of installation you are planning. For hollow wood partitions use toggle-bolts. If you cut into a plaster wall, there are boxes which have expanding and tightening sides for a firm, snug fit. Normally, inside the boxes, plan for not more than one No.

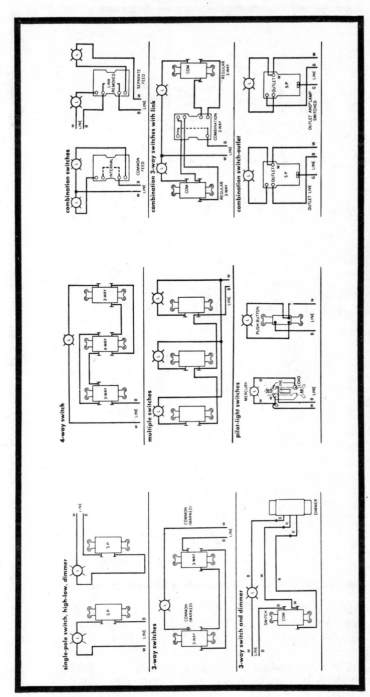

Fig. 8-33. Some important switch wiring diagrams.

163

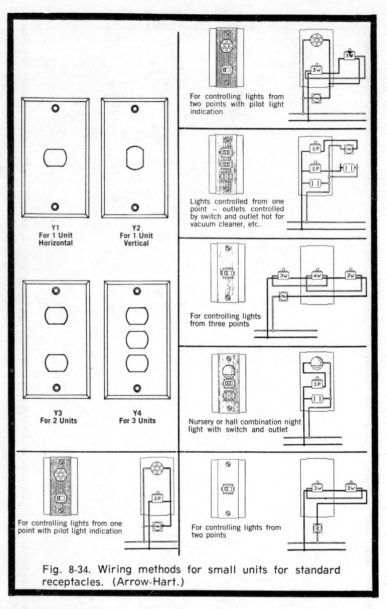

For controlling lights from two points with pilot light indication

Lights controlled from one point — outlets controlled by switch and outlet hot for vacuum cleaner, etc.

For controlling lights from three points

Nursery or hall combination night light with switch and outlet

For controlling lights from one point with pilot light indication

For controlling lights from two points

Y1
For 1 Unit
Horizontal

Y2
For 1 Unit
Vertical

Y3
For 2 Units

Y4
For 3 Units

Fig. 8-34. Wiring methods for small units for standard receptacles. (Arrow-Hart.)

14 wire for each 2 cu in. of space, or 2 ¼ cu in. for each No. 12 wire. This means that for three No. 12 conductors, the box dimensions should be at least 1 x 2 x 3 in. Most boxes are larger than this.

Flexible conduit is hollow and, like solid tubing, has to have the wiring pulled through it **after** it is installed.

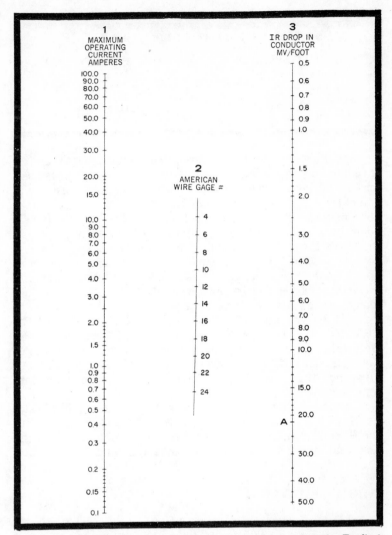

Fig. 8-35. Voltage loss vs distance and wire size nomograph. To find **maximum current carrying capacity** recommended for any standard wire size, use a straightedge to connect from the wire size on scale 2 to point A (for example) on scale 3, then read max current on scale 1. To find **voltage drop in millivolts per foot** for known wire size and operating current, connect the known current on scale 1 and the wire size on scale 2, and read voltage drop on scale 3. To find **wire size required** for known operating current and known maximum tolerable voltage drop across supply leads, determine maximum tolerable drop in millivolts per foot of wire (sum of **two** leads), then connect the value now known on scale 3 to the know current on scale 1. Read wire size on scale 2. Based on an arbitrary minimum 500 circular mils per ampere. High-temperature-class insulation will safely allow higher currents.

Table 8-1. Support of Rigid Conduit	
CONDUIT SIZE, IN.	**SUPPORT SPACING, FT**
½	4
¾	4
1 to 2½	4 to 5
2½ to 3	6

Speaking of pulling wires through a conduit, there is a kind of reasonably rigid steel tape called a "fishline" which is used for this purpose. Here's how it's used: You simply insert one end of the "fishline" into the conduit at the box opening, then, bending a small hook in the end of the "fishline" or "fishtape," hook the wires to it carefully, and from the starting box **pull** the wires through the conduit. Sometimes you have to add a powder or special grease (which your dealer can tell you about) if the run is long, or has a few bends (remember no more than four 90 degrees or a total of 360 degrees of bending along one run between boxes), or if you have several wires to pull. ALSO you pull **ALL WIRES AT ONE TIME, NEVER TRY TO PULL THEM THROUGH ONE AT A TIME.** Determine ahead of time how many wires you will need, get the right size conduit for this number of wires, and pull them all at the same time. Table 8-1 gives the support requirements for common sizes of conduit.

In installations where you go through or under the joists, use flexible conduit. Going through the center of the joists with the flexible conduit eliminates the requirement to notch the joist and install a "cover board" over the conduit in the notches.

SAFETY (THE CODE)

It is important to remember that where exposed or concealed wiring conductors in tubes or cables are installed through bored holes in studs, joists, or similar wood members, the holes **shall** be bored at the approximate center of the wood members or at least 2 in. from the edge. If there is no objection by the city inspectors, metal clad cable, or nonmetallic sheathed cable, or aluminum sheathed (note: this is NOT **aluminum cable conductors**) and type "MI" cable may be laid in notches, in studs, or joists **when covered by a metal plate** 1/16 in. thick to prevent nails being driven into the cables when the surface boards are in place and the cabling is concealed.

Cables of one or more conductors buried in the earth **shall** be of a type approved for the purpose and use. Such types as "USE" and "UF" may be used. Where single-conductor cables are installed, all conductors of each service, feeder, subfeeder, or branch circuit including the neutral conductor **shall** be run continuously in the same trench or raceway. Supplemental mechanical protection such as covering board, concrete pad, raceway, etc. may be required by your own city inspector. An example of raceways in a concrete base is shown in Fig. 8-36.

You may think that we've used too much emphasis and boldfacing of certain words above and to follow. Just remember: Forewarned is forearmed.

The above information is useful if you intend to have outside wiring in the yard or if you plan runs from the house to the garage (if it is separate), or to the barn or other outside buildings. In each of these cases, we highly recommend that the house or "hot" end of the line be fused **in all wires** in such a manner that all can quickly be disconnected. A circuit breaker or a switch type fuse panel is suitable.

Flexible **nonmetallic tubing** as a raceway for your wires may be used in dry locations and where it will not be exposed to severe physical damage. The tubing **shall** be in continuous lengths not exceeding 15 ft for each (and these may then be connected together by suitable fittings) and the tubing **shall** be secured to the surface by straps spaced not to exceed 4.5 ft apart. If metal clad type "MC" cable is used, this **must** be secured every 6 ft and within 2 ft of a box or outlet.

BEFORE YOU BUILD OR BUY A HOUSE

If you are building or buying a home, here are some electrical concepts which you need to check:

a. In every kitchen, family room, dining room, breakfast room, living room, parlor, library, den, sun room, recreation room, and bedroom, the receptacle outlets **shall** be placed so that no point along the floor line **in any wall shall** be more than

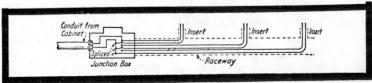

Fig. 8-36. A method of running conduit in a concrete floor pad. The junction box is used for all taps, and each line to an outlet or fixture must be intact without other taps, and run through its own length of conduit from the junction box.

6 ft, measured horizontally, from an outlet in that space, including any wall space 2 ft wide or greater, and the wall space occupied by sliding panels in exterior walls. The wall space afforded by fixed dividers such as free standing bar type counters shall be included in the 6 ft measurement.

b. In the kitchen and dining areas, a receptacle outlet shall be installed at each counter space wider than 12 in. Counter spaces separated by range tops, refrigerators, and sinks shall be considered as separate counter top spaces. Receptacles rendered inaccessible by the installation of stationary appliances will not be considered as satisfying these minimum outlet requirements. All required outlets must be accessible at all times. Receptacle outlets shall, insofar as possible, be spaced equal distances apart. Receptacle outlets in floors shall not be considered or counted as part of the required number of wall receptacle outlets unless located close to the wall. At least one wall type receptacle outlet shall be located in the bathroom close to the basin location.

c. Outlets in other sections of the dwelling for special appliances such as laundry locations shall be placed within 6 ft of the intended location of the appliance (garbage disposals, refrigerators, ovens, etc., are under this category). At least one wall receptacle shall be installed for the laundry.

d. For residential occupancies, all 120V, single phase 15A and 20A receptacle outlets installed outdoors shall have approved ground-fault circuit protection for personnel. Such ground-fault circuit protection may be provided for other circuits and locations and occupancies, and where used will provide additional protection against the line-to-ground shock hazard. This type fuse opens the circuit if it becomes grounded or if a danger exists due to leakage current.

GROUND FAULT INDICATOR-PROTECTOR

Harvey Hubbell Co. now has a model (GFP-115) that can be plugged into any dual outlet wired for the 3-wire system. The outlet must be rated at least 15A, 117V. This is the "standard-trip" rating of this model, as a circuit breaker. The trip-rating as a ground fault indicator-interrupter (GFI) is 5.0 mA. Differential transformers and a solid-state amplifier are used. It is simply plugged into one of the receptacles; a captive screw holds it in place and automatically makes the safety-ground connection. This goes into the center hole between the outlets.

Any appliance of not more than 15A rating can then be used in safety from the GFI outlet. A TEST button is provided.

When this is pushed, the pilot light should come on, indicating that the unit is working. Pushing the RESET button sets it for use again. Of course, if there is any leakage in whatever you plug into it—drill, saw, extension cord, etc.—the GFI will trip, indicating that the device is not safe to operate. You must then get it repaired or determine how the short is being made in the device and eliminate the cause.

We deliberately omit schematics and pictures of ground fault indicators-detectors-interrupters because of their complexity—only a licensed electrician should install them!

Publication ASA-C91.1 is the American Standard Requirement for Residential Wiring. You might locate it through your local library for more detailed study of practices and procedures.

Special Wiring Systems and Concepts

In this chapter we explain some of the fancier electrical systems; those in which you have remote control over devices, timers on devices or circuits, heating of driveways or roof tops, etc. It will be a broad chapter and in it you may find just the concept or circuit you need.

SWITCHES

We start with an example of simply running an electric line from inside the house to the outside—two wires which must be connected to a "knife-type" switch, that, in turn, may have a fuse socket in each leg. You can buy this kind of switch so that when the knife handle is closed, the fuses are in the circuit, and when the knife switch is open you have completely disconnected **both** lines from the inside electric power lines completely and totally. This is a dandy device to use on the farm or in a yard where pole lights are necessary. The switching can be done from the house end of the line. Figure 9-1A shows the proper way to connect it electrically and Fig. 9-2 shows how to physically install the knife or blade type switch. Be sure you read the notes on these drawings. One thing we always liked about this type of switch is that all you have to do **is look at it** and see if the circuits are connected or not. No guesswork here! One bad feature, of course, is that the blades are exposed to anyone who might touch them, thus care is required to mount the switch high above the reach of children (and pets) and out of the way so even inexperienced adults won't accidentally get across the blades or terminals.

THERMOSTATS AND TIMERS

Let's now discuss two kinds of switches often used: the temperature controlled switch (TCS) and time clock switches. The TCS is used on air conditioners and in home heating and cooling systems as a **thermostat**. It can also control attic fans.

The time clock switch is used for sprinkler systems, for lights, for breakfast coffee, for burglar prevention, etc.

While we were visiting friends one summer, their refrigerated air conditioner went on the blink. The drive motor burned up. We replaced it with some difficulty, and then the new motor began acting peculiarly when we set the temperature control on a lower than high position. The motor acted as if it had a much higher than normal load. It tended to slow down. When we put the control on HIGH it speeded up and was okay. We had the new motor burn out also before we found

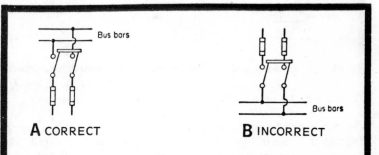

A CORRECT   **B** INCORRECT

Fig. 9-1. Method of wiring of an open knife switch. The bus bars are the regular house line. Note that in A, when the switch is open, the **blades** on the switch are disconnected from the line. The situation on the right (B) is incorrect, as this keeps the blades hot.

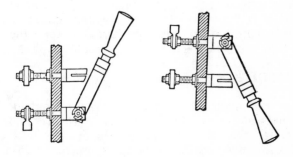

Fig. 9-2. The knife switch should always be mounted so that the pull of gravity will tend to OPEN the contacts.

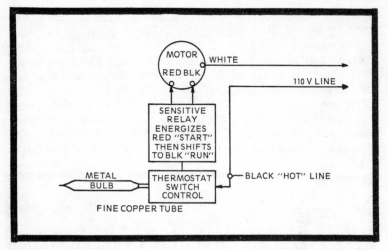

Fig. 9-3. A general wiring schematic for a refrigerated air conditioner. The metal bulb is usually located just behind the room side screen.

the trouble to be in the little thermostat; a sealed unit placed to measure the temperature in the room side of the unit. The thermostat control is about the size of a lead pencil and has a small copper line attached to it. It causes a switch to be **closed** by high temperature, Fig. 9-3, and opened when the temperature is low or cold. We traced the circuit and found the main control switch also was pitted and blackened where overload of current, due to the malfunctioning of the thermostat, had wreaked its havoc. Replacement of both units, **the thermostat and the control switch**, was the remedy. The story is recited here because one does not usually suspect that little unit as causing trouble. But the thermostat is present in EVERY window type refrigerated air conditioning unit. **Suspect it** if the motor tends to become draggy, slow in operation. Turn the power off, quickly, and repair the unit or, as we did, you'll buy some expensive motors.

AIR CONDITIONING

The roof mounted air conditioner is an outside device which uses electricity and which must be wired safely. We have stated that normally all conductors are in a conduit which is grounded, so there are no exposed wires. But inside these units, the evaporative type especially, there are pumps which one may replace, and leads to the blower motor itself **and these that are not usually placed in conduit.** Rather, they are merely extension cords that have plugs on them which, in turn, plug into a receptacle socket. This terminates the conduit inside the cooler housing. Examine one of these cords and you'll find that because the wire is subject to dampness and wet water spray, it is moldy, and the plugs and fittings are rusted and have become dangerous. Replace them. You should see to it that the new wires are of an approved type for the installation concerned, that adequate protection is taken for the receptacle and plugs so that a minimum of corrosion takes place. ALWAYS COMPLETELY DISCONNECT THE WIRING THROUGH THE FUSE BOX AND ANY SWITCHES to be sure there are no hot wires in the air conditioner **before** you attempt **any adjustment**, replacement, or change to the wires or components of this unit! We have always found it best to turn on the machine at its inside switch, then go to the fuse panel and set the **circuit breaker** there to **off**, listening and

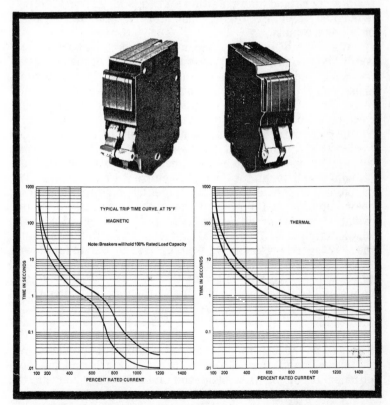

Fig. 9-4. The magnetic type circuit breaker and operating characteristic curves. Notice that on the left curve, if you have a 200 percent overload, the circuit breaker will trip in 10 to 30 seconds (the distance between the two lines).

watching to observe when the air conditioner unit STOPS running. Then turn off the unit's inside switch and **then** make the replacements. Even so, be CAREFUL! Some characteristics of a circuit breaker switch such as are found on fuse panels is shown in Fig. 9-4.

One other note about the air conditioner. In winter time, especially at Christmas time, this unit can furnish a source of outside electrical power for lights around the top of the house. Just plug the light string into the receptacle in this unit, using care not to have any wires placed that will fray or be damaged by the sides or fittings, and then you can operate your Christmas lights from the inside air conditioner switch. By the way, use only miniature lights on man-made trees; larger bulbs generate too much heat and cause pockets to form on the branches. For metallic trees, use off-the-tree spot or floodlights rather than traditional strings of lights; a voltage leak caused by faulty wiring could cause a fatal electric shock or a severe burn.

We wanted to change the motor in our evaporative air conditioner so that we had two speeds instead of the one speed for which the unit was originally wired. Two problems presented themselves. First, the new motor had color coded terminals which we did not understand; and second, there were only three wires in the conduit from the roof receptacle to the two switches inside the house. One of these switches turned the air conditioner **blower motor** on and the second turned the water pump motor on. We could not run an additional line through the conduit because it was too small. The problem was solved by asking an electrician about the motor color code and it follows the house code perfectly. The terminal with a **white dot** is the COMMON, the second with the **black dot** is the HIGH and the third terminal with the **red dot** is the LOW speed for the motor switch. Next we had to find a way to get the proper switching with just the two switches and three wires. The solution was to use the switch originally connected to the water pump line to turn on the water pump **and** the blower motor on LOW speed simultaneously. Then if the second switch was turned on this should change **only** the **blower speed** to HIGH. How was it done? Examine the circuit, Fig. 9-5. If you use this circuit, **please** install the relay in a **waterproof** and **weatherproof** case OUTSIDE the air conditioner. If you install it **inside** the air conditioner housing, the moisture might play havoc with the relay contacts and wiring.

In sum, this is a method of using only three wires, and two off-on switches to get HIGH and LOW motor speeds and keep the pump running when the blower is on. The relay is a 110V ac single pole double throw type.

It is important when using electric motors to have some idea of the size of their starting currents and running currents. We show this in Table 9-1. With a knowledge of these currents, you can select the proper wire size to run to the motor, the proper switch size (in volts and amperes) to switch the motors off and on, and select the proper relay switch so its contacts will be rated at slightly more than the starting currents so excess pitting and burning of contacts will not occur.

Speaking of relays, if you use them, make sure to inspect them once in a while to see that they are clean and have the proper spacing. If they start making bad connections due to dirt or oxidization, then you may have a "hot" contact. This can also occur on switches of every type and kind. When it happens, the resistance to the flow of electricity through the switch or contact of a relay will cause more heating and more resistance, etc., until a burnout occurs. Keeping relay con-

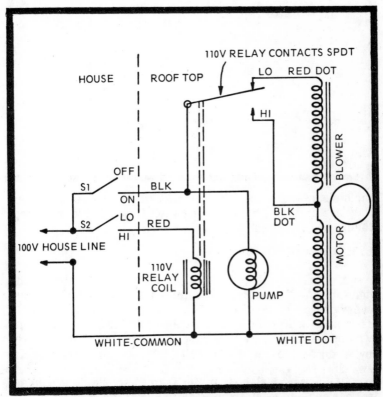

Fig. 9-5. A relay can be used to give Hi-Lo motor control as well as control over the pump motor of an air conditioner. Switches S1 and S2 are located in the house and the three wires, red, black, and white, are run in conduit up to the rooftop-type air conditioner unit.

Table 9-1. Starting and Running Currents of Motors

MOTOR HP	VOLTS	STARTING CURRENT, AMPERES	LOAD CURRENT, AMPERES
1/6	120	15	4.4
1/4	120	20	5
1/3	120	20	5.5
1/2	120	22	7
3/4	120	28	9.5
1/4	240	10	2.5
1/3	240	10	3
1/2	240	11	3.5
3/4	240	14	4.7
	240	16	5.5
1 1/2	240	22	7.6

tacts clean and burnished, and checking switches to be sure they are making good contact (when it is possible to inspect them) helps prevent trouble.

SPECIAL CIRCUITS

A couple of other circuits are nice to know about. The light-operated (photocell) switch can be purchased most everywhere. It comes in two types, one which uses a resistance solid-state cell and transistors and relay, and one which is a heavy-duty type which costs more but it activated by a solid-state resistance cell and a mechanical thermal switch.

Of course, all electric motor frames should be grounded, so that you can't get a shock by touching the frame itself. If you need more than the normal 120V supply, then **call an electrician** for a 220 or 440V installation. No matter how tempted, NEVER use extension cords to run an electric motor. That would be like playing Russian roulette!

The first type is normally used to turn on a lamp **in the house** at sundown; the second, which is weatherproof, obtainable from Sears, Roebuck, may be installed **outside** to turn on a yard light at night and off in the daytime automatically. The difference between these two types is that the second type is sealed and extremely rugged. It is more expensive, and it is more reliable, as it does not rely on a relay, as such. It has a metallic strip inside, around which is wound a resistance wire, and this is connected in series with the light cell.

The first type mentioned is simply connected to an outlet receptacle by means of its own plug, and whatever is to be controlled is then connected to the unit itself through its receptacle. Some units are designed to be screwed into a lamp socket and the light bulb screws into them. A nice convenient solution for the nighttime lighting problem.

It seems we have digressed somewhat from our discussion of thermostatic controlled units. Let us return to that basic idea. In Fig. 9-6 is an example of the very common house unit,

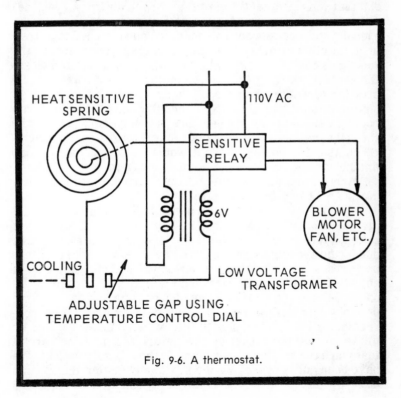

Fig. 9-6. A thermostat.

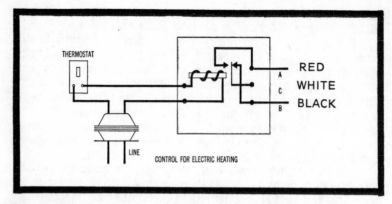

Fig. 9-7. This diagram shows how a thermostat controlled relay is used. Here, the relay's single pole double throw contacts replace the on-off switch which could be placed at points A, B, and C. Notice how the relay will energize the line C-B in the position shown, and when energized will disconnect this line and connect lines C-A. (G.E.)

which is found mounted on the wall and which has a rotating dial that you use to set the temperature you desire. As you see, in the figure, when you move this dial you actually close or open the spacing between two contacts, one fastened to a fixed arm, the other attached to a heat-sensitive spring. Thus, as the room gets colder, for example, the spring contracts and closes the contact, starting the blower motor in the furnace or the attic fan or whatever. If the unit has a contact on the opposite side of the spring level (dotted), then this can be used to start a cooling unit when the temperature gets high enough to make the coil expand to such an extent that it makes contact there. Of course, the cooling contact would be connected through another sensitive relay to another blower motor.

Another example of the electrical schematic for a thermostat is shown in Fig. 9-7.

The timer, or time clock, is a dandy device to have around the house or office. It is used at home to start the morning coffee, or to turn room lamps off and on at various times during the evening when you are away. The type used for these purposes is a simple, small, motor-driven unit obtainable from any electric parts house for about $5. Simply set the two arms on the device, one for off and one for on time and set the time properly by rotating the dial. Then plug the unit into a convenient wall socket and plug the unit which is to be operated by the timer into the socket on the timer. You must be careful that you have set the proper arms to the proper positions. We have been guilty of making the mistake of setting the on time at 5—we thought this was to be 5 **PM**, and instead set it to 5

AM. This is easy to do! Be careful. Also, on some units is either a slide type or rotating type switch which permits you to check to see if current is passing into the timer. You turn it, or slide it, and a light which is supposed to be off is turned on. But now you must be careful when resetting the switch that you don't turn it too far, or get it into an improper setting. Care on this adjustment will prevent trouble. By the way, if the lamp is supposed to be off, there is a chance you'll plug it in and forget to turn in on. This must be done. We always recommend turning the lamp switch on, THEN pull its plug and insert it into the timer socket. You can see that it was on as it is supposed to be. If the timer is properly set, the light will go off. When leaving home for a few days, it is wise to have several of these units, set at different ON and OFF times to switch lamps on in the house. We have also used a timer to turn a radio on and off at hight in about the same time frame as we normally operate the TV. Anyone coming to the house to check it would hear the music, see the lights, and possibly be hesitant about breaking in.

The heavier duty timers can be used to turn the sprinkler systems on and off late at night. This is a money-saving idea, since the evaporative effect of the sun is minimized and better watering is obtained in this manner. One must obtain an electrically operated solenoid valve (or several of them) from your electrical dealers. Get the operating instructions, install the unit at the water line outside, and here, of course, you must be certain to have weatherproof electrical wiring and connections. Normally these are run in conduit and we suggest you have a qualified electrician do this for you. Then you set the timer as you choose for the watering period you desire and it will operate the valves to water the yard without further attention from you.

When you obtain a timer it is best to spend a little more money and get a good one. You simply double the expense by trying to save with a cheap, inferior unit. Get one which is rugged, has a specification on it that it **can** handle the wattage you plan to use it for, and which has a UL (Underwriters Lab) approval on it. If, for example, during the Christmas season you plan to have one to turn the Christmas house lights on and off, then be sure to add up the wattage. Bulbs normally are 7.5W **each** so multiply this times the number of bulbs, say 50, and you'll find you need a timer capable of handling 400 to 600W for this job alone. Good, heavy-duty contacts are necessary, so read the specifications and be sure that the one you get can do the job. Of course, if you get one which can handle, say, 500W, it can handle any wattage below that without even trying!

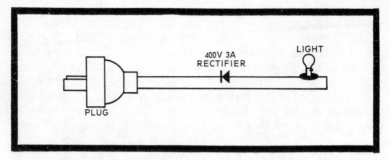

Fig. 9-8. A method of extending the life of the ordinary lamp bulb. Any small rectifier can be used. They can be found in radio parts stores.

Since we are speaking of lights, there is a good little circuit, Fig. 9-8, which will add months and months of life to an outside light. You simply use a little solid-state rectifier in series with the line to the bulb. We have found these to be small enough to be physically placed in the sockets themselves in most cases. What the rectifier does is to cause the current through the light to go in only one direction, direct current, as opposed to the bidirectional flow (120 times per second) of the normal house current. This system cuts light output drastically. It should be used with a higher-wattage bulb than normal; this will help compensate for the power drop resulting from rectification.

HEATERS

Now let us consider some concepts of space heating with electricity. There are electrical heaters which can be built into walls or around the baseboard and which provide nice, even, clean heat. There is a general formula which relates the amount of wattage necessary to the size of the room to be heated. It is:

$$\text{Watts required} = (1)(\text{volume of air space})$$

Let's calculate a simple example. If you have a room which measures 10 x 10 x 8 ft, then the volume is the product of these numbers or 800 cu ft. Since the wattage is equal to the volume, we will need at least 800W to properly heat the room. The heating elements or heaters should be distributed uniformly around the room walls or baseboards for even heating of this air space. Figure 9-9 shows one such installation. The heaters should be arranged around the room for uniform heat

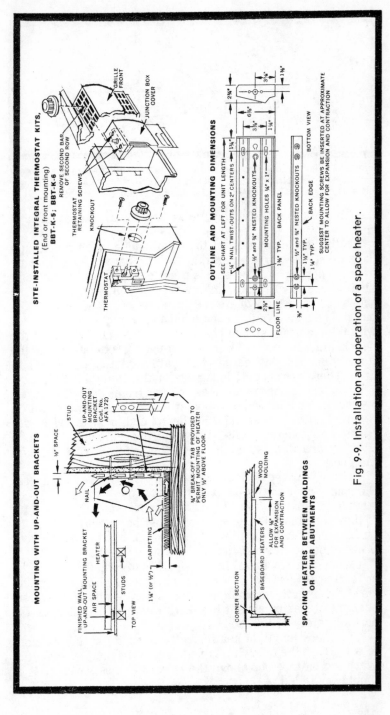

Fig. 9-9. Installation and operation of a space heater.

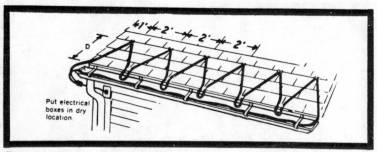

Put electrical boxes in dry location.

Fig. 9-10. Electric heating cords are available in most large department stores, at the electric counter. Here is a method of installation for winter time. During the summer, the cord would be removed and stored.

distribution and preferably be at low level, as heat always rises.

If the heating panels are to be embedded in the walls, then care must be exercised so that the temperature will not be over 150 degrees F. The panels must **NOT** be mounted on any kind of wood or flammable wall, but only on plaster or sheetrock unless adequate insulation is provided around the panel. The panels must be installed as complete units together with their wiring, and no changes to the wiring lengths of the panels must be made. The panels must not extend beyond the room which is to be heated—for example, into a closet or hall area. These are not safe locations for this kind of installation. The panels must be at least 8 in. from doors, fixtures, switches, receptacles, or pipes, or outlet junction boxes in a room. Heating cables **shall not** be spliced or installed in dry board and plaster walls. They can be embedded only in Gypsum board, plaster lath, and other walls of fire-resistant material.

There is a type of electrical heating wire made by General Electric which can be affixed to roofs, or to walks or pipes outside to cause melting of snow and ice formations,(Fig. 9-10). If these are installed, great care must be taken to follow instructions exactly to prevent the possibility of fire to your office or home, or short circuits in ground installations.

When using switches for these kinds of circuits, the switch or thermostat or whatever device controls the flow of current, **must** open **all** ungrounded conductors; and of course, adequate fuses or circuit breakers must be installed as the normal safety precaution.

To check length of heating cable needed for roof deicing, select "D" in the following table to equal the number of feet of overhang on the roof above the eave. For best results, ½ ft should be added.

D = 1 ft—use 1.8 x length of roof edge for length of cable needed for zig-zag part of installation.

D = 2 ft—use 2.6 x length of roof edge for zig-zag length.

D = 3 ft—use 3.5 x length of roof edge for zig-zag length.

Add to the above, the length of the return cable plus the additional footage required to reach the receptacle.

Example: If length of roof is 40 ft and D needs to be 1 ft, then 40 x 1.8 = 72 ft of zig-zag. Now add the 40 ft return through the gutter plus 6 ft of lead to reach the receptacle (located 3 ft from roof edge). Total heating length needed thus becomes 72 + 40 + 6 = 118 ft. Use one 60 ft cable **loop** (120 ft total length) for the installation.

When using this kind of heating wire to prevent pipe freeze-ups, you need to use the graph in Fig. 9-11. Here are the instructions on how to use the graph: From the minimum ambient temperature on the graph (which should be the lowest outdoor temperature on record for the area), draw a line up to the curved line representing the size of pipe or hose to be protected. Then from this point, draw a horizontal line (either left or right) to the "WATTS PER FOOT OR PIPE." For example: 4 in. pipe, minimum temperature—minus 25 degrees F. Follow dashed line from minus 25 degrees F up to its intersection with the 4 in. pipe line, then over to the right to find 12.1 as the watts required per foot of pipe using 1 in. of approved insulation and paper wrap. For installation where

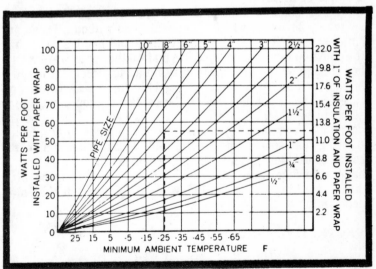

Fig. 9-11. The heating electric cord can be used to prevent pipe freeze-up. This graph shows the watts per foot required for various size pipes based on the minimum temperature desired.

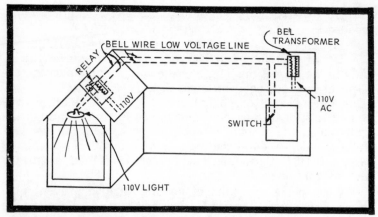

Fig. 9-12. A method of running a low voltage line through the attic to energize a 120V light on the garage door. The switch can be located at the opposite end of the house.

only paper wrap is used, move left from the intersection point to find 55 watts required per foot of pipe.

Once in a while the electric **water heater** in the house will have its element shorted electrically, especially if the area in which you live has a high chemical content in the water. When this happens the fuse panel circuit breaker trips, or a fuse blows, and the element must be replaced. **This is a job for the electric company!**

On the surface of the tank, there are two small thermostat switches, however, which you should know about. These are small screwdriver-set shafts, usually found just inside each of the doors on the water heater. If you find that your water is not hot enough or too hot, these may be adjusted clockwise to a higher turn off level, or counterclockwise, to a lower turn off level. Sometimes they have to be replaced even though they seem to be okay. The replacement is required because the constant water heat changes the "springiness" of the activating leaf spring and so the unit doesn't function in the range of temperatures it should. Even for these, it is best to have a service man attend to it. But we include this little discussion here to let you know some things on it that may need fixing.

PROTECTIVE CIRCUITS

There are some other circuits which are useful around the home or office. One is the protection of your automobile by having lights on the garage as shown in Fig. 9-12. Here a low

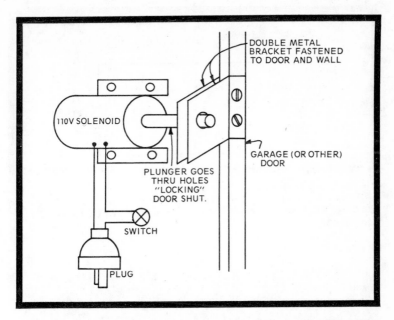

Fig. 9-13. Use of a solenoid with a plunger to make a garage (or other) electric door lock. When energized, the plunger retracts, allowing the door to be opened.

Fig. 9-14. The G.E. mercury button can provide a contact when tilted.

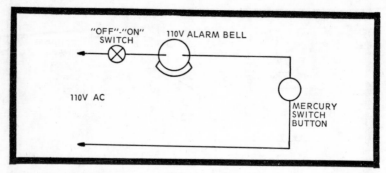

Fig. 9-15. How the mercury switch, or a normally open contact type microswitch can be used. The microswitch contacts will close if the door is moved slightly so the lever arm of this type switch can no longer contact the door.

Fig. 9-16. A contact type switch. Pressure against the "button" can keep the switch contacts open or closed. There are two types of this switch.

voltage is used to operate a relay which then turns the lights on. A small relay permits the operation of the 120V bulb from a long distance without having to use the higher voltage line over the greater distances.

Did you know that you can obtain a solenoid which has a protruding metal piece armature which can be used to fit into a hole in a metal bracket on the garage door, so that you must activate a switch to get the door open? You have probably run into such a device in a bank or other similar place where someone pushes a button to activate a device which permits a door to be opened. What happens is that the armature or plunger, which operates against a spring, is pulled back into the coil of wire when the coil is energized. Ask your dealer about this kind of safety device. Figure 9-13 shows a circuit.

Another way to use a protection circuit is to obtain a mercury switch (Fig. 9-14). Mount this on the door so that if the door is lifted or tilted the mercury will roll and make contact. This in turn can be connected to an alarm bell. Thus, whenever the garage door is raised, and if the circuit is complete to the mercury switch, an alarm will sound. Figure 9-15 shows how to wire such a circuit.

There are various other techniques which will give an intruder a moment's panic. One is to use a microswitch, available at most radio parts stores, which can be caused to close a contact when the garage door is moved from the position which normally holds this kind of switch open. See Fig. 9-16.

Another method is to use a photoelectric cell and a light. This system is also available at most radio parts stores. Just place a mirror on the door to be guarded. Shine the light at the mirror and position the mirror so that it reflects the light into the photoelectric cell. The photocell will have a relay connected to it so that when an interruption to the light occurs, the relay will close and an alarm will be sounded. With this system, just a slight movement of a door (or window) will cause the alarm bell or horn to sound. These are complete units selling for about $20, and they are very reliable.

Now we examine a burglar alarm schematic in Fig. 9-17. Note how all switches, for example, are connected in parallel so that opening any window will close one of them and cause the alarm bell to ring. This system can be operated on a low voltage ac circuit or batteries may be used. However, if batteries are used, be sure they have the life to operate over a long time. Use large sizes and check them often to be sure they do not deteriorate. In the circuit shown the battery can be replaced with a standard bell ringing transformer to make the system all-electric.

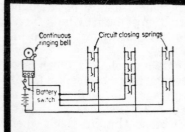

Fig. 9-17. A method of wiring windows with normally open switches which are arranged so that the least movement of the window, or door, will cause the contacts to close and sound the alarm. No current is drawn unless an intruder attempts to enter.

A home signaling circuit is shown in Fig. 9-18. This is desired, for example, in homes where elderly people are housed or invalids are cared for. The bells can be located near the person who would use them. Always remember that older people are weak. **Do not** use a button switch or any other kind **which requires much strength** to operate! Get the easiest to operate switch that you can find, such as the GE mercury switch or a microswitch. Some attractive pushbutton switches are shown in Fig. 9-19.

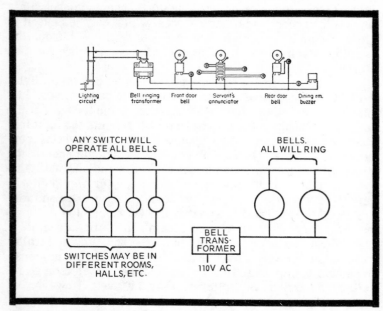

Fig. 9-18. Two circuits which can be used to summon people. The first uses individual buttons and bells; it is selective. The second circuit rings bells in many locations with any switch. This is a good alarm type system.

Fig. 9-19. Bell type button switches can be most attractive. They can also be inconspicuous when blended into woordwork as ornaments.

REMOTE CONTROL SWITCHING

There are so many times that we want to control lights or other devices by remote lines, and there are so many cases where we already have this situation, in thermostatic controls, etc., that a few minutes looking over the concepts of low voltage remote control switching is worthwhile. The following units are made by General Electric, and well illustrate the

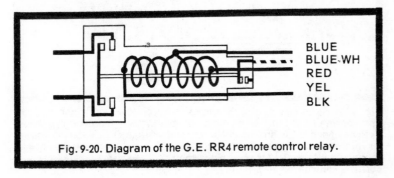

BLUE
BLUE-WH
RED
YEL
BLK

Fig. 9-20. Diagram of the G.E. RR4 remote control relay.

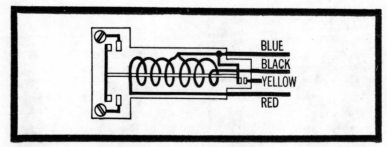

Fig. 9-21. The internal wiring of the G.E. RR8 relay is similar to the RR4, but the current required is less. This, like the RR4, is a 2-position relay. When a momentary contact is made to the winding, it will energize and stay energized. To deenergize it, a second momentary contact must be made to the release winding.

concepts and methods. Some of the advantages of this kind of control (using a low voltage to operate a relay, which, in turn, operates the lights or motors or other devices) are as follows:

a. You can have any choice of switching arrangements you desire through the use of direct switches or motor controlled switches.

b. You have control from multiple locations and through the use of motor controlled switching you can control a variety of functions simply and positively.

c. There is maximum safety because you switch at low voltages. Cost is reduced because you do not have to run lines insulated for high voltages. In many cases, ordinary bell wire can be used for control switching.

d. You can have master off-on control of multiple protection lights from a single switch.

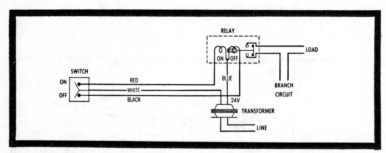

Fig. 9-22. Basic circuit of G.E. remote control relay switching system. The normally open, single-pole double-throw, momentary-contact switches permit the small control current to flow ONLY during the interval that a switch is pressed. Therefore, any number of switches can be connected in parallel for multipoint control. The split-coil relay permits **positive control** for on and off.

e. There are lockout circuits for extra protection from or for children.

f. Locator switches that glow in the dark can be used when they are hidden, and pilot switches show when lights are on when they are in hidden areas and not easily checked.

Let's examine the basic units of this kind of circuit wiring. The relays are shown in Figs. 9-20 and 9-21. The relay has a set of contacts at the bottom, which control the 120 or 220V electric lines. These two wires are connected in series with the power plug and the unit to be operated. The control wires are at the top. A momentary contact between the blue and black leads will cause the armature to pull up and close the contacts. There is another connection (RED) at the top which when energized momentarily will cause the relay to open its contacts. A third lead, the blue-white one, will energize a pilot relay when you desire to use this.

Next let use examine a basic circuit using the relay (Fig. 9-22. This kind of relay requires only a small amount of current to flow during the time the relay is energized. It consumes no current to remain closed or open. To open the contacts, the second winding is energized momentarily and this opens the contacts.

All wiring for the following is color coded and the code is shown in Fig. 9-23.

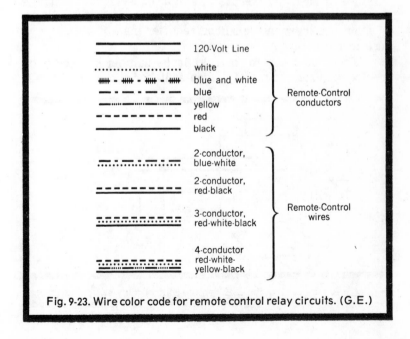

Fig. 9-23. Wire color code for remote control relay circuits. (G.E.)

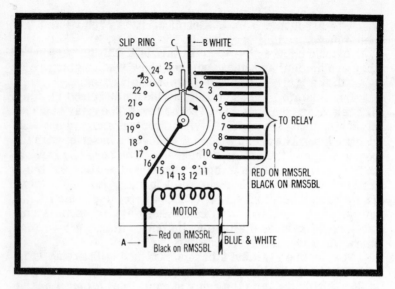

Fig. 9-24. Internal wiring of the RMS5BL or RMS5RL master motor control unit. (G.E.)

Although G.E. relays will operate satisfactorily on standard alternating current as it comes from a 24V transformer, the use of half-wave-rectified, single-phase 24V ac permits better performance and affords greater protection to the system. An inexpensive silicon rectifier may be obtained for use in the diagrams to follow.

When you desire to control a large number of circuits by remote control, the motor driven switcher shown in Fig. 9-24

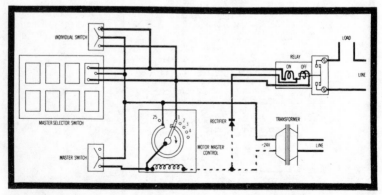

Fig. 9-25. Basic circuit of the G.E. remote control relay switching system. Note the 3-position individual switch which is spring loaded to a middle **off** position.

would be used. And a basic circuit using the relays and the motor driven switch is shown in Fig. 9-25.

The way you would operate this is to select the individual circuits you want to energize by depressing the local MASTER SELECTOR switch and then you would simply push the MASTER SWITCH and this will cause all of those circuits which you have designated to be either on or off, to be energized in the proper manner to accomplish your desires.

In Fig. 9-24, the motor-master control unit starting switch is connected to lead A and white lead B. When the starting switch is activated, contact arm C starts to rotate in the direction indicated. The arm connects the white lead to contact points 1 through 25. It also picks up the slip ring connected to the white lead and keeps it operating until one full revolution is completed. Complete revolution takes 17 seconds.

Figure 9-26 shows another application which may be of value to you. Here is shown the wiring to connect the units and a timer or photocell.

Notice that the timer of photocell (marked P.E. cell) simply turns the master unit on at some predetermined time, or upon the reception of light, or darkness as you designate. Once the master unit is on, the motor master control then begins operation and will activate, or inactivate as you have designated, as many as 25 other circuits. Other type operational programs are possible with these master control units, relays, and switches.

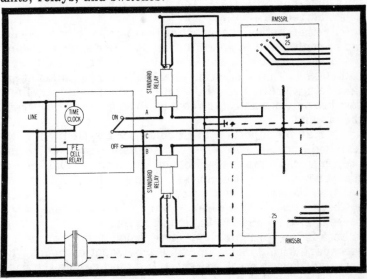

Fig. 9-26. Use of the master motor control with a clock timer or a photocell relay control circuit. (G.E.)

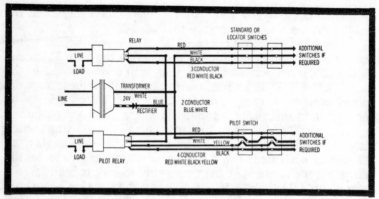

Fig. 9-27. The basic circuit of the remote control wiring units. The transformer, rectifier, relays, and bidirectional contact switches are shown.

A basic circuit for remote control operation of this low voltage system is shown in Fig. 9-27. Notice that the power transformer is in the center and the operating switches are to the right. Any number of switches may be used so long as they are the two-contact type. Remember you must make a momentary contact to energize a relay, and another momentary contact to deenergize a relay. The relays are to the left on each side of the transformer and the load (whatever devices you want operated, lights, etc.) would be connected at the points marked LOAD. The **line** is the connection to the house current of 120V ac.

Another method of connecting the time clock to the master motor control unit is shown in Fig. 9-28. Here you will notice the use of a manual control switch connected across the timer output lines. This will permit checking the installation at any time to be sure it will all operate. Note also that the timer supplies only a momentary contact to energize or deenergize the circuits.

It may be important to troubleshoot a system once it is installed and there is a very systematic approach to this problem. Also there are symptoms which you should know so that you will be able to diagnose troubles easier.

1. **If the system won't work:**

a. Check for power at the transformer. Check to see that 24V is present **at the output** of the transformer.

b. Look for an open circuit or short circuit in the transformer secondary wiring. Sometimes the movement of running the wires will cause a break or a short. If a short circuit, there will be heating of the transformer and you won't be able to measure any voltage at the terminals of the secondary.

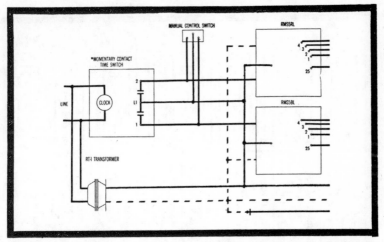

Fig. 9-28. Since only a momentary contact is necessary to energize or deenergize a circuit, the clock timer need supply only this type contact. Note the manual control override for test purposes or to energize circuits at times other than provided for by the timer.

c. You may have to use an isolation procedure to locate which branch circuit is shorted. The best way to do this is to disconnect all branches, then reconnect them one at a time to find out which one, when connected, produces the failure of all.

d. Check to see if all switches work. Some may not make contact, or some may be shorted or the connections to them might be shorted. You may have to connect a "test switch" across each or in place of each switch to be tested.

2. **If an individual circuit stays on and won't turn off,** this fault is caused by a continuously energized relay coil. The following procedure then should be used:

a. Check to see if any individual switch is shorted or the connections to it are shorted. It may be that a physical mounting problem exists which has caused the switch to be energized and just remounting the unit will clear the trouble.

b. If the control switch is suspected, then remove the wiring from the switch and test by touching the wires together (at 24V, this can be done without danger) to see if the circuit energizes and deenergizes. If this happens, then the switch may be faulty.

c. Inspect the relay for a possible short circuit between its connections at the low voltage end. Sometimes solder flows too much and causes a short here.

d. If the relay just won't operate and yet everything seems okay, then suspect the relay itself. Check it directly across the transformer terminals and see if it energizes and deenergizes.

RATING	RECEPTACLE WIRING	RATING	RECEPTACLE WIRING	RATING	RECEPTACLE WIRING
15A 125V 2-POLE 2-WIRE NEMA #1-15R Cat. No. GE4025 Page E17	125V NEUTRAL WHITE	50A 250V 2-POLE 3-WIRE NEMA #6-50R Cat. No. GE4141 Page E26	250V EQUIPMENT GROUND GREEN	30A 125/250V 3-POLE 4-WIRE NEMA #14-30R Cat. No. GE4191 Page E24	250V 125V 125V NEUTRAL WHITE EQUIPMENT GROUND GREEN
20A 250V 2-POLE 2-WIRE NEMA #2-20R Cat. No. GE4120 Page E20	250V	15A 277V 2-POLE 3-WIRE NEMA #7-15R Cat. No. GE4056 Page E11	277V NEUTRAL WHITE EQUIPMENT GROUND GREEN	50A 125/250V 3-POLE 4-WIRE NEMA #14-50R Cat. No. GE4181 Page E27	250V 125V 125V NEUTRAL WHITE EQUIPMENT GROUND GREEN
15A 125V 2-POLE 3-WIRE NEMA #5-15R Cat. No. GE4065 Page E4	125V NEUTRAL WHITE EQUIPMENT GROUND GREEN	20A 277V 2-POLE 3-WIRE NEMA #7-20R Cat. No. GE4290 Page E20	277V NEUTRAL WHITE EQUIPMENT GROUND GREEN	60A 125/250V 3-POLE 4-WIRE NEMA #14-60R Cat. No. GE4171 Page E28	250V 125V 125V NEUTRAL WHITE EQUIPMENT GROUND GREEN
20A 125V 2-POLE 3-WIRE NEMA #5-20R Cat. No. GE4108 Page E12	125V NEUTRAL WHITE EQUIPMENT GROUND GREEN	30A 277V 2-POLE 3-WIRE NEMA #7-30R Cat. No. GE4235 Page E24	277V EQUIPMENT GROUND GREEN	20A 250V 3-POLE 4-WIRE NEMA #15-20R Cat. No. GE4200 Page E21	250V 250V 250V EQUIPMENT GROUND GREEN
30A 125V 2-POLE 3-WIRE NEMA #5-30R Cat. No. GE4138 Page E23	125V NEUTRAL WHITE EQUIPMENT GROUND GREEN	50A 277V 2-POLE 3-WIRE NEMA #7-50R Cat. No. GE4255 Page E27	277V EQUIPMENT GROUND GREEN	30A 250V 3-POLE 4-WIRE NEMA #15-30R Cat. No. GE4230 Page E24	250V 250V 250V EQUIPMENT GROUND GREEN
50A 125V 2-POLE 3-WIRE NEMA #5-50R Cat. No. GE4140 Page E25	125V NEUTRAL WHITE EQUIPMENT GROUND GREEN	20A 125/250V 3-POLE 3-WIRE NEMA #10-20R Cat. No. GE4124 Page E20	125V 250V 125V NEUTRAL WHITE	50A 250V 3-POLE 4-WIRE NEMA #15-50R Cat. No. GE4250 Page E28	250V 250V 250V EQUIPMENT GROUND GREEN
15A 250V 2-POLE 3-WIRE NEMA #6-15R Cat. No. GE4067 Page E9	250V EQUIPMENT GROUND GREEN	30A 125/250V 3-POLE 3-WIRE NEMA #10-30R Cat. No. GE4132 Page E22	125V 250V 125V NEUTRAL WHITE	60A 250V 3-POLE 4-WIRE NEMA #15-60R Cat. No. GE4260 Page E28	250V 250V 250V EQUIPMENT GROUND GREEN
20A 250V 2-POLE 3-WIRE NEMA #6-20R Cat. No. GE4188 Page E15	250V EQUIPMENT GROUND GREEN	50A 125/250V 3-POLE 3-WIRE NEMA #10-50R Cat. No. GE4152 Page E25	125V 250V 125V NEUTRAL WHITE	20A 120/208V 4-POLE 4-WIRE NEMA #18-20R Cat. No. GE4127 Page E21	120V 208V 208V 120V NEUTRAL WHITE
30A 250V 2-POLE 3-WIRE	250V	20A 125/250V 3-POLE 4-WIRE	250V	60A 125/250V 4-POLE 4-WIRE	250V

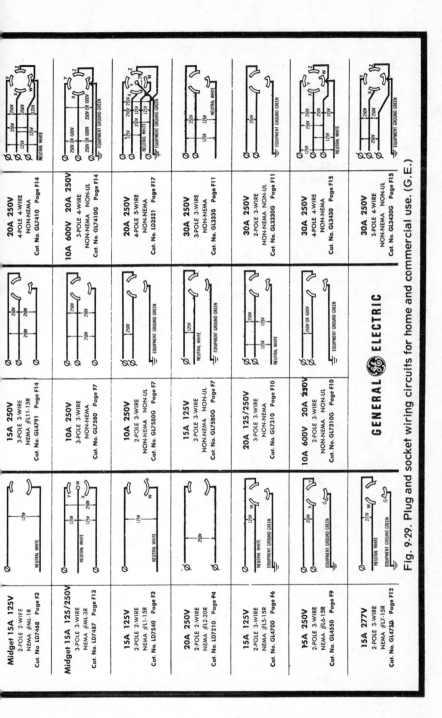

Fig. 9-29. Plug and socket wiring circuits for home and commercial use. (G.E.)

197

e. If a circuit turns on when it should be off, your wiring is reversed at the control switch. If you use the color code supplied by the manufacturer, you won't have this difficulty.

For further information, of course, G.E. Wiring Device Dept. at Providence, Rhode Island 12904, will be glad to help out.

So we now have some additional ideas for adding internal and external circuits and controls to our lighting and electrical outlet system. We believe these can be of value to you in your planning, construction, repair, or operation of electrical things around your house.

PLUGS & SOCKETS

We had promised, in an earlier chapter, that we would show you some of the various types of plugs and socket connections for large heavy-duty appliances. We want to keep that promise now and refer you to Fig. 9-29. It could be that sometime you may have need for this information and here it is in handy form for your use.

Bibliography

1. Electrical Wiring Interiors
   Graham   Am. Technical Society

2. Electrician - Study Guide for Civil Service and License
                        Exams
   Liebers   U.S. Govt.

3. Electricity - Study and Teaching
                        U.S. Govt.

4. 1972 National Electrical Code

5. 10th Edition National Electrical Code
   Abbott and Stetka

6. IES Handbook

7. Electrical manufacturers pamphlets: General Electric, GTE Sylvania, Emerson Electric Co., Loran, Inc., etc. Available from distributors. See Yellow Pages of telephone book.

8. Outdoor Lighting
   Holophane Co., Inc., 1120 Avenue of Americas
                        New York, N.Y. 10036

9. Fluorescent Lighting Association
                        250 E. 43rd St.
                        New York, N.Y. 10017

10. Recessed and Surface Fixtures
    Kirlin Co., 3435 E. Jefferson Ave.
    Detroit, Mich. 48207

11. Rambusch Lighting Co.
    40 West 13th St.
    New York, N.Y. 10011

12. Stonco Electric Products Co.
    Kenilworth, New Jersey 07033

13. Container Corporation of America
    (Ask for Color Harmony Manual)
    38 South Dearborn St.
    Chicago, Illinois

14. Your local library—for your local electrical code, etc.

E

F

G

H

I

K

L